First Hitting Time Regression Models

To my children:

Mark, Helen and Stephen

Mathematical Models and Methods in Reliability Set
coordinated by
Nikolaos Limnios and Bo Henry Lindqvist

Volume 4

First Hitting Time Regression Models

Lifetime Data Analysis Based on Underlying Stochastic Processes

Chrysseis Caroni

iSTE

WILEY

First published 2017 in Great Britain and the United States by ISTE Ltd and John Wiley & Sons, Inc.

ISTE Ltd
27-37 St George's Road
London SW19 4EU
UK

www.iste.co.uk

John Wiley & Sons, Inc.
111 River Street
Hoboken, NJ 07030
USA

www.wiley.com

Library of Congress Control Number: 2017938379

British Library Cataloguing-in-Publication Data
A CIP record for this book is available from the British Library
ISBN 978-1-84821-889-5

Contents

Preface

The analysis of lifetime (or time-to-event) data is one of the busiest fields of statistics. This is partly due to the interesting theoretical issues that come up, which tend to be rather different from those that are seen in other areas, and, above all, due to the enormous range of its practical applications. One major field of application is in engineering and technology, where the subject is usually known as reliability. Reliability is concerned with how long a machine or a machine component will operate before it requires replacement or repair. The other major field of application is in biostatistics. A classic example of this field of application is the study of patients' survival after therapy. Other important fields of application include finance and banking, as well as other areas.

In biostatistics, information about the individuals in the study may be available in the form of covariates (e.g. age and gender). Exploiting this information may improve prediction of lifetimes (a younger patient will generally do better than an older one). Covariates are usually included in statistical models in the form of regression models. Lifetime data analysis contains a rich range of regression models, which will be presented in this book. Many biostatistical texts restrict their attention almost exclusively to a famous regression model introduced by D.R. Cox; this is the model that predominates in statistical practice. However, there is much to be said for using models that are not simply empirical (as Cox's model is) but are substantive in the sense that they provide a description of the underlying process. The emphasis of this book is on models of this type, in particular the one based on the concept that a lifetime ends when an underlying stochastic process reaches a boundary for the first time.

It is a pleasure to acknowledge my debt to several people who helped to inspire my interest in lifetime data analysis, through their own work, their willingness to discuss matters with me and respond to my queries, and their invitations to participate in conferences and visit their institutions. In this respect, I wish to record my gratitude especially to Mei-Ling Ting Lee, Alex Whitmore and Nikolaos Limnios. In addition, I thank Alan Kimber and Martin Crowder who first got me interested in lifetime data. Finally, Polychronis Economou and Dimitris Stogiannis, first as doctoral students and then as collaborators, have aided my work in many ways.

Chrysseis CARONI
May 2017

1

Introduction to Lifetime Data and Regression Models

1.1. Basics

The principal aim of this book is to present a particular approach to the regression analysis of lifetime data in detail, and to discuss its features and advantages in comparison to older-established approaches. This objective will first involve undertaking a general review of the various regression methods that can be found in the literature.

This opening chapter provides a brief review of the basic features of lifetime data and its modeling. First of all, what is meant by lifetime data? A great deal of statisticians' activity goes into studying whether or not an event occurs and how various factors influence its occurrence. If we move on from the simple yes/no fact of occurrence to also examining how long it takes until the event occurs, we enter the realm of "time to event" or "lifetime" data. Basic examples include how long a machine operates until it breaks down (the event is the breakdown) and how long a patient lives after undergoing heart transplantation (the event is death). These examples show two major areas of application. One is in engineering and technology (where the subject is usually known as reliability modeling) and the other is in biomedical sciences (known as survival analysis). However, other areas of application include all those in which statistics is used - in other words, in virtually every science. As the two examples of a machine's breakdown and a patient's survival suggest, applications of lifetime data analysis can have immense practical importance. Well-known textbooks with wide coverage of lifetime data analysis include

Lawless [LAW 03], Collett [COL 14], Kalbfleisch and Prentice [KAL 02] and Klein and Moeschberger [KLE 03]. A brief review was given by Hougaard [HOU 99]. Other papers reviewing the analysis of lifetime data include Kiefer [KIE 88] in economics and Chung *et al.* [CHU 91] in criminology.

In mathematical terms, lifetime is denoted by T. No two machines are identical or operate under identical conditions; no two people are quite alike. Consequently, we treat T as a random variable, which follows some distribution in the relevant population of machines or people (in general, units). We note that T must be non-negative. Furthermore, in this book, we will follow the vast majority of the literature in treating the time scale as continuous. Consequently, we suppose that $T \sim f(t), t > 0$ for some probability density function (pdf) $f(.)$, and hence that $F(t) = P(T \leq t) = \int_0^t f(u)du$. The functions that present particular interest are the following:

– the survival function $S(t) = \bar{F}(t) = 1 - F(t) = P(T > t) = \int_t^\infty f(u)du$

– the hazard function $h(t) = f(t)/S(t)$.

The former is $P(T > t)$, the probability of survival for at least time t - the probability that the machine is still operating, or that the patient is still alive, after this time. In engineering and technological applications, this probability is called reliability and the notation $R(t)$ is usually used instead of $S(t)$. The term hazard function is replaced by failure rate. Other terms in use for the same function include force of mortality (in demography) and intensity.

The survival function or reliability $P(T > t)$ is a quantity of basic scientific and practical importance. For example, in medical settings, a patient's prognosis might be expressed as his or her five-year survival probability, and in manufacturing, reliability is obviously related to how long a guarantee period can be offered for a product. The hazard function can be interpreted as the instantaneous rate of failure at time t, given that the unit has survived that long, and hence the term failure rate. However, it is important to remember that the hazard refers to failure conditionally on survival to that time (the unconditional failure rate is of course given by the pdf of the lifetime distribution). More precisely, the hazard function gives the conditional probability of failure in the next short interval of time $(t, t + \delta t]$, for a unit that is still functioning at time t:

$$P(t < T \leq t + \delta t | T > t) = h(t)\delta t. \qquad [1.1]$$

Also useful and important is the cumulative hazard, $H(t) = \int_0^t h(u)du = -\ln S(t)$ as well as the mean residual life, given by

$$\mu(t) = \frac{\int_t^\infty S(u)du}{S(t)},$$

which is the expected lifetime still to come for a unit that has already survived until time t.

The functions f, F, S, h and H are all equivalent, in the sense that knowing any one of them enables all the others to be deduced. Complete details can be found in standard references (e.g. Lawless [LAW 03]), which also present detailed expressions for the more widely used parametric lifetime distributions $f(.)$, such as the exponential, Weibull, gamma, log-normal and others. In the following section, we present the best known parametric lifetime distribution as an example, the Weibull distribution. Some details of another parametric distribution - the inverse Gaussian distribution - can be found in section 2.5 and elsewhere in the text. Some general properties of lifetime distributions are presented briefly by Olkin [OLK 16] and at length by Marshall and Olkin [MAR 07].

Some aspects of these distributions that have major importance in the analysis of lifetime data, such as the hazard function, present little interest in other fields of statistics. Conversely, some properties of distributions that have great general importance do not concern us much in lifetime data analysis. The prime example is the mean of the distribution. This is because most lifetime distributions are highly skew, with a long tail to the right. For a distribution of this shape, the median is usually reported rather than the mean. Furthermore, in practice, the restricted duration of a study may make it difficult to estimate the mean accurately (see comments in section 1.9). However, sometimes a restricted mean survival time (RMST) can be used. By the definition of the mean μ of a distribution, and assuming that it exists, we have

$$\mu = E(T) = \int_0^\infty tf(t)dt = \int_0^\infty S(t)dt.$$

Adapting this, we define the RMST up to time t^* as

$$\mu(t^*) = \int_0^{t^*} S(t)dt.$$

This can also be written as $\mu(t^*) = E[\min(T, t^*)]$. The RMST thus represents the expected duration of survival up to time t^*: in other words, how much of the interval $(0, t^*)$ an individual will survive, on average. RMSTs can be computed for different values of t^* and compared between groups of subjects; this will be especially useful if the relation between survival in the groups is not simple (e.g. the first group does better than the second initially, but later on, the second group has the lower hazard). See for example A'hern [AHE 16] and references therein.

Two further general comments about lifetime data must be made. First, it is a characteristic feature of such data that not all the units under study will actually experience the event during the study. Some patients will still be alive when the medical researcher closes the data file for analysis; some machines will still be functioning when the time allotted to the study runs out. The lifetimes of these units are said to be right censored at the times when they were observed. They provide information that must be taken into account in the analysis even though this information takes the different form $T > t$ rather than $T = t$. This can only be done easily if the censoring process is uninformative about the lifetime (see section 3.1). Other types of censoring as well as the less common phenomenon of truncation are discussed in standard references (see, for example, Lawless [LAW 03, Chapter 2]).

The second additional comment is the observation that a "lifetime" need not correspond to clock time, or even be measured in units of time at all. For a machine, the relevant time may be the time for which it is actually operating, excluding periods when it is turned off or is idle. For a car, the operating "time" would probably be measured better by how many kilometers it has covered rather than by the calendar age of the vehicle, because this will be the more important factor as far as wear and tear is concerned. Sometimes there may be several alternatives: for an aircraft, for example, calendar age, flight hours and number of landings could all be relevant measures of lifetime (see Duchesne and Lawless [DUC 00]). The question of the appropriate time scale is also discussed by Farewell and Cox [FAR 79], Oakes [OAK 95] and Kordonsky and Gertsbakh [KOR 93, KOR 97] as well as others. Later on, we will see cases where overall "time" is a weighted sum of the durations of the periods of time spent in different states (e.g. the movement of an employee through different jobs with varying exposure to health risks).

Finally, we observe that the concept of a non-negative random variable describing the point at which an event occurs can be adapted to cases where the variable is not a time at all, but, for example, the load placed on a structure. The load is increased until the structure fails.

1.2. The classic lifetime distribution: the Weibull distribution

Here, for the purpose of illustration, we provide details of the *Weibull distribution* (named after the Swede Waloddi Weibull), which is the most widely used parametric model for lifetime data. Empirically, it has been found to fit well to data of many kinds, and in fact, its use with lifetime data can be justified by theoretical arguments (see below).

The pdf of the Weibull distribution in one common parameterization is

$$f(t) = \exp\{-(t/\alpha)^\eta\}\eta t^{\eta-1}/\alpha^\eta, \quad t > 0,$$

where $\alpha > 0$ is the scale parameter and $\eta > 0$ is the shape parameter. The special case $\eta = 1$ gives the exponential distribution. The survival or reliability function is

$$\begin{aligned} S(t) &= \int_t^\infty e^{-(u/\alpha)^\eta} \eta\alpha^{-1}(u/\alpha)^{\eta-1} du \\ &= \int_{(t/\alpha)^\eta}^\infty e^{-v} dv \quad [\text{substituting } v = (u/\alpha)^\eta] \\ &= \left[-e^{-v}\right]_{(t/\alpha)^\eta}^\infty \\ &= \exp\{-(t/\alpha)^\eta\} \end{aligned}$$

and therefore the hazard function is

$$h(t) = \eta\, t^{\eta-1}/\alpha^\eta, \quad t > 0.$$

The behavior of the hazard function is as follows:

$$h(t) = \begin{cases} increasing, & \text{when } \eta > 1 \\ constant, & \text{when } \eta = 1 \\ decreasing, & \text{when } \eta < 1. \end{cases}$$

This means that the Weibull distribution is quite flexible when it comes to describing lifetime data. However, it is unable to capture various features that are sometimes observed in hazard functions in real life, such as when the hazard increases to a peak and then falls, or when it falls to a minimum and then increases.

Figure 1.1 presents examples of the shapes of the Weibull pdf, survival function and hazard function for various values of the parameters of the distribution. Note that the distribution is skewed to the right, which is a characteristic feature of lifetime distributions.

The expected value and the variance of the lifetime T can be found using the following expression for the r-th moment of the distribution:

$$\begin{aligned} E(T^r) &= \int_0^\infty t^r f(t) dt \\ &= \int_0^\infty \alpha^r u^{r/\eta} e^{-u} du \quad [\text{substituting } u = (t/\alpha)^\eta] \\ &= \alpha^r \int_0^\infty u^{r/\eta} e^{-u} du \\ &= \alpha^r \Gamma(1 + r/\eta), \end{aligned}$$

where $\Gamma(\cdot)$ is the gamma function. Setting $r = 1$ and 2 gives

$$E(T) = \alpha \Gamma(1 + \eta^{-1}) \quad \text{and} \quad E(T^2) = \alpha^2 \Gamma(1 + 2\eta^{-1})$$

and hence, the variance of the lifetime T is

$$V(T) = \alpha^2 \left[\Gamma(1 + 2\eta^{-1}) - \{\Gamma(1 + \eta^{-1})\}^2 \right].$$

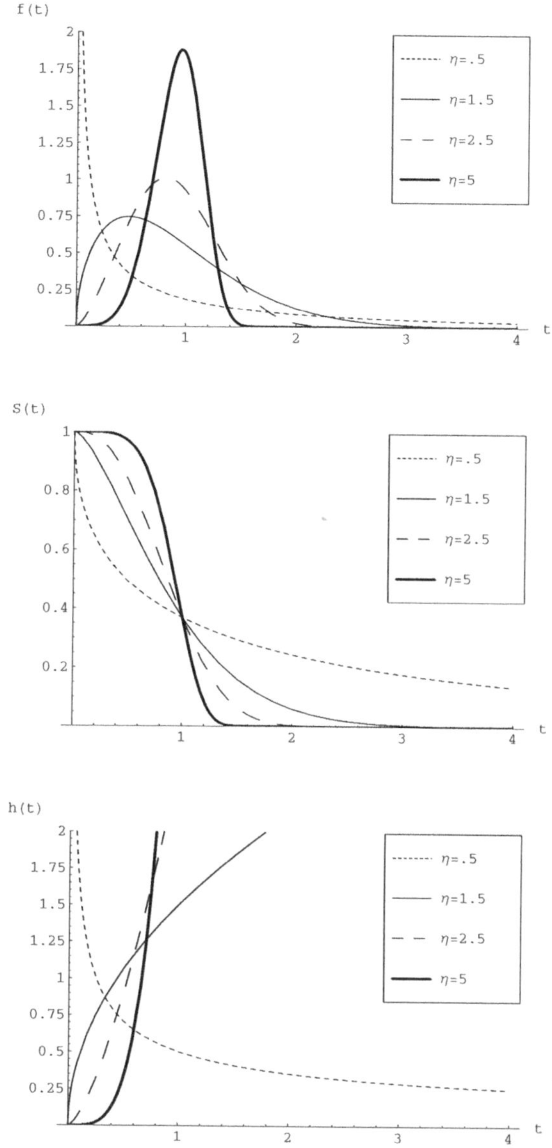

Figure 1.1. *Plots of the pdf (upper diagram), survival function (middle diagram) and hazard function (lower diagram) of the Weibull distribution for selected values of* η, *with* $\alpha = 1$

The following alternative parameterization of the Weibull distribution is also often seen in the literature:

$$f(t) = \alpha\beta t^{\beta-1}e^{-\alpha t^{\beta}},\ t > 0,\ \alpha > 0,\ \beta > 0.$$

Countless examples of applications of the Weibull distribution can be found in the literature, which mostly concern the lifetimes or strengths of materials. One application of a different kind is by McDonald *et al.* [MCD 96] to the lifetimes of a species of bird. By fitting a Weibull distribution to lifetimes and finding that the shape parameter is greater than one, they concluded that mortality rates increase with age - so-called actuarial senescence. This was claimed to be the first demonstration of the phenomenon in an unmanipulated, natural population and thus constituted the first empirical evidence against a long-held assumption that mortality of birds is generally independent of age.

The theoretical argument for the Weibull distribution's widespread use in practical situations is the following "weakest link" argument. Many of the units that we study can be regarded as being made up of smaller components or parts, and it may be reasonable to suppose that the durability or strength of the whole is equal to the durability or strength of the weakest part, just as a chain is made up of links and the chain's strength is given by the strength of its weakest link. Given this structure, the distribution of the unit's lifetime is determined by the distribution of the *minimum* of the set of random variables that represent the lifetimes of the unit's components. Statistical theory demonstrates that only certain distributions have the necessary properties to represent such a minimum. One of these *extreme value distributions* is the Weibull.

The literature contains many lifetime distributions, most of which do not see any practical application. For example, many extensions of the Weibull distribution have been devised (see, for example, Caroni [CAR 14a]). In gaining extra flexibility, these extensions lose appealing properties of the Weibull distribution, such as the extreme value interpretation and the properties of the regression models that will be discussed later on in this chapter.

1.3. Regression models for lifetimes

Although the fact that no two units can be identical means that there will always be a random component in the lifetime, in part it may be possible to predict the lifetime from the factors or covariates that describe the unit or the conditions under which it has been operating. A patient's prognosis after an operation, for example, is likely to depend to some degree on his or her age, on medical history and on the variables that describe the state of health at the time of the operation. An older patient, in poor condition and with a long history of ill health will be expected to have a shorter time-to-event (death, relapse) than a younger patient who started out in better shape. Car tires would be expected to wear out quicker if the vehicle is often driven off-road.

The concept of introducing the dependence of an outcome variable such as the duration of a lifetime on the values of covariates is familiar from the multiple linear regression model

$$y = \boldsymbol{\beta}'\mathbf{x} + \epsilon,$$

where $\mathbf{x} = (x_0, x_1, \dots, x_p)'$ is the vector of covariates, with $x_0 \equiv 1$. The standard model takes the distribution of the random error term as $\epsilon \sim N(0, \sigma^2)$, in which case the model for the dependent variable can be written as

$$y \sim N(\mu, \sigma^2) \text{ with } \mu = \mu(\mathbf{x}) = \boldsymbol{\beta}'\mathbf{x}. \qquad [1.2]$$

This expression suggests one way of extending a regression model to situations where it is not reasonable to assume a normally distributed dependent variable: select a more appropriate distribution (e.g. Poisson with parameter μ) and link its parameters in some way to the linear predictor $\boldsymbol{\beta}'\mathbf{x}$ formed from the covariates (e.g. $\ln \mu = \boldsymbol{\beta}'\mathbf{x}$ is often an appropriate choice in combination with the Poisson distribution). In this way, we obtain the class of generalized linear models (GLM) in which the mean parameter is related to the linear predictor (see McCullagh and Nelder [MCC 89]). For example, the GLM version of [1.2] when the dependent variable Y is a count of the number of events and therefore might follow the Poisson distribution is

$$y \sim \text{Poisson } (\mu) \text{ with } \ln \mu = \boldsymbol{\beta}'\mathbf{x}.$$

In the much wider class of generalized additive models for location, scale and shape (GAMLSS), as many as four parameters of the distribution can depend on covariates (see Rigby and Stasinopoulos [RIG 05]).

The same approach can be taken to distributions that are often used in modeling lifetime data. For example, the inverse Gaussian distribution and the gamma distribution both belong to the exponential family that is modeled in the standard framework of generalized linear models. The inverse Gaussian distribution will be mentioned in this context in section 3.2. However, there are other ways of approaching the matter in the context of lifetime data, which give rise to the general classes of models that will be considered in the following sections.

Parametric lifetime regression models are usually fitted by direct maximization of a likelihood function using numerical methods. Given a sample of n independent observations $\{(t_i, \mathbf{x}_i, \delta_i), i = 1, \ldots, n\}$, where unit i with covariates $\mathbf{x}_i$ has lifetime t_i and censoring indicator δ_i (=1 if t_i is an observed failure time, 0 if t_i is a right censored observation time), the likelihood is

$$L(\boldsymbol{\theta}) = \prod_{i=1}^{n} f(t_i|\mathbf{x}_i, \boldsymbol{\theta})^{\delta_i} S(t_i|\mathbf{x}_i, \boldsymbol{\theta})^{1-\delta_i}, \qquad [1.3]$$

where the parameter vector $\boldsymbol{\theta}$ includes the regression coefficients. Using the relationships between the probability density function, hazard function and survival function, this likelihood can be written in various alternative forms, if desired. For example, using $h(t) = f(t)/S(t)$ to substitute for $f(t)$, the likelihood can be written in terms of the hazard and survival functions as

$$L(\boldsymbol{\theta}) = \prod_{i=1}^{n} h(t_i|\mathbf{x}_i, \boldsymbol{\theta})^{\delta_i} S(t_i|\mathbf{x}_i, \boldsymbol{\theta}),$$

which may sometimes be convenient.

1.4. Proportional hazards models

To illustrate one of the main approaches to the regression modeling of lifetime data, we begin with the widely used Weibull distribution. Note that this does not fall within the framework of generalized linear models. Its survival function is

$$S(t) = \exp\{-(t/\alpha)^\eta\}, \;\; t > 0, \; \alpha > 0, \; \eta > 0.$$

We introduce the effect of covariates $\mathbf{x}$ on the parameters of the model, giving

$$S(t|\mathbf{x}) = \exp\{-(t/\alpha(\mathbf{x}))^\eta\}.$$

The scale parameter α now depends on $\mathbf{x}$. (This is the usual form of model, although it is possible to allow the parameter η to depend on $\mathbf{x}$ instead, or to let both the parameters depend on covariates at the same time. In the latter case, the covariates affecting α and η do not need to be the same. A recent paper by Burke and MacKenzie discusses the general approach where both the parameters depend on the covariates for the Weibull distribution and in general [BUR 16a]. See section 3.2.)

Let $\alpha(\mathbf{x}) = \alpha e^{\boldsymbol{\beta}'\mathbf{x}}$ or simply $\alpha(\mathbf{x}) = e^{\boldsymbol{\beta}'\mathbf{x}}$ since the constant α can be absorbed into the exponent. (Once again, this is the usual form of the model, although not the only possibility.) Note that the function $e^{\boldsymbol{\beta}'\mathbf{x}}$ is positive, a restriction that is necessary here.

The hazard function is readily obtained from $h(t) = -\frac{d}{dt}\ln S(t)$ as

$$h(t|\mathbf{x}) = \eta t^{\eta-1} e^{\boldsymbol{\theta}'\mathbf{x}},$$

where $\boldsymbol{\theta} = -\eta\boldsymbol{\beta}$. Now compare the hazard functions of the two units with covariate vectors $\mathbf{x}_1$ and $\mathbf{x}_2$. Their ratio is

$$\frac{h(t|\mathbf{x}_1)}{h(t|\mathbf{x}_2)} = \frac{\eta t^{\eta-1} e^{\boldsymbol{\theta}'\mathbf{x}_1}}{\eta t^{\eta-1} e^{\boldsymbol{\theta}'\mathbf{x}_2}} = e^{\boldsymbol{\theta}'(\mathbf{x}_1-\mathbf{x}_2)},$$

which does not depend on time. In other words, the hazard function of one unit remains in constant proportion to the hazard of the other. This is the proportional hazards (PH) model, which applies for any non-negative $\alpha(\mathbf{x})$, not just $e^{\boldsymbol{\beta}'\mathbf{x}}$. A particular version of the PH model - Cox's semi-parametric PH regression model - has virtually become the standard model for evaluating biomedical lifetime data. This model will be described in section 1.14.

Another example of a model that possesses the PH property is obtained from the Gompertz distribution, which is described most simply by its hazard function

$$h(t) = \mu\phi^t,\ \mu > 0,$$

which is decreasing in t for $\phi < 1$, increasing for $\phi > 1$ and constant for $\phi = 1$, in which case it reduces to the exponential distribution. The PH model modifies the distribution into another Gompertz distribution with a different value of μ but the same ϕ (see Hougaard [HOU 99]). The use of the Gompertz distribution is restricted mainly to demography and actuarial science, where it has a long history. It has been used, for example, to describe mortality among adults. In these contexts, $\phi > 1$ (increasing hazard - i.e. mortality - at older ages).

Now consider what happens if $\phi < 1$. In the survival function

$$S(t) = \exp[-\mu\,(\phi^t - 1)\,/\ln\phi]$$

the term ϕ^t tends to zero, therefore the limit of $S(t)$ as t tends to infinity is not zero. In fact, with $\mu > 0$ as before, the limit of $S(t)$ is e^ξ where $\xi = \mu/\ln\phi < 0$ and hence $0 < S(\infty) < 1$. This version of the Gompertz distribution is called the negative Gompertz distribution by Marshall and Olkin [MAR 07, Chapter 10]). Because $S(\infty) < 1$, it is an improper distribution or defective distribution. However, this feature is not necessarily a defect as far as using the distribution as a statistical model goes. The existence of a positive probability mass at infinity can be interpreted to mean that the corresponding proportion of the population will never die. This assumption is clearly meaningless in the actuarial study of human mortality, but could possibly be very realistic in a shorter-term study of mortality from a disease after a treatment. In the latter case, those who "never die" (at least, from the disease under study) are those who have been cured of the disease. Thus, the apparent defect becomes an asset of the model in its ability to model data. Examples of the application of the Gompertz distribution that exploit this characteristic include Cantor and Shuster [CAN 92] and Gieser *et al.* [GIE 98].

This feature will be mentioned (under the name of cured fraction or long-term survivors) quite often in this book, because it is shared by the inverse Gaussian distribution, which, as will appear in due course, is a central topic.

Both the basic examples of a distribution that possess the PH property, the Weibull and the Gompertz distributions, have hazard functions that are monotonic in t. This is not a necessary condition for a PH distribution. If we write the PH property in its general form

$$h(t|\mathbf{x}) = h_0(t)g(\mathbf{x}),$$

where $h_0(t)$ is a baseline hazard function, it is obvious that if the baseline hazard function $h_0(t)$ has a maximum or a minimum at a value t_0, then the hazard functions $h(t|\mathbf{x})$ for every $\mathbf{x}$ likewise have maxima or minima as the case may be at this same value t_0. For example, if the hazard function falls to a minimum and thereafter increases - often claimed to be a realistic form in various situations - then that minimum would have to occur at the same time irrespective of the values of the covariates. This seems unlikely to be true in practice.

Bagdonavičius and Nikulin [BAG 99] proposed an extension of the PH model to the generalized PH model. The hazard function can be written as

$$h(t|\mathbf{x}) = r\left\{\mathbf{x}(t)\right\} q\left\{H(t|\mathbf{x})\right\} h_0(t),$$

where r and q are positive functions. Thus, the hazard rate at time t depends not only on the current values of the covariates (as in PH) but also on their history as expressed by the cumulative hazard $H(t)$. One special case is the generalized linear PH model, in which $r(\mathbf{x}) = e^{\beta'\mathbf{x}}$ as usual and

$$q\left\{H(t|\mathbf{x})\right\} = e^{\gamma H(t|\mathbf{x})},$$

so that

$$h(t|\mathbf{x}) = e^{\beta'\mathbf{x}+\gamma H(t|\mathbf{x})} h_0(t).$$

Thus, the cumulative hazard up to this moment in time is treated as an additional, unknown covariate. This model is examined further by Bagdonavičius *et al.* [BAG 05].

1.5. Checking the proportional hazards assumption

The theory that was outlined in the preceding paragraphs requires the assumption of PH. If this assumption is inappropriate for the data, then it is meaningless to fit this particular regression model. How can we check that the assumption is appropriate?

The hazard function

$$h(t|\mathbf{x}) = h_0(t)e^{\boldsymbol{\beta}'\mathbf{x}}$$

gives the survival function

$$S(t|\mathbf{x}) = \exp\{-H_0(t)e^{\boldsymbol{\beta}'\mathbf{x}}\},$$

where $H_0(t)$ is the cumulative hazard function corresponding to the baseline hazard function $h_0(t)$.

Consequently,

$$\ln\{-\ln S(t|\mathbf{x})\} - \ln H_0(t) = \boldsymbol{\beta}'\mathbf{x}$$

which means that the curves

$$\ln\{-\ln S(t|\mathbf{x})\}$$

for different values of $\mathbf{x}$ are simply the horizontally displaced versions of the curve $\ln H_0(t)$ when plotted against t. Consequently, all the curves $\ln\{-\ln S(t|\mathbf{x}_i)\}$ for different $\mathbf{x}_i$ are parallel to each other.

This observation suggests a simple way of checking for PH:

– compute non-parametric Kaplan-Meier estimates of the survivor function $\hat{S}(t|\mathbf{x})$ for selected $\mathbf{x}$;

– plot $\ln\{-\ln \hat{S}(t|\mathbf{x})\}$ against t for each selected $\mathbf{x}$.

If all the lines for the various $\mathbf{x}$ are indeed parallel to each other, then the assumption of the proportional hazards is correct. This idea applies to *any* PH model, but does not tell us which distribution is the appropriate one if we are to carry out a parametric regression. Also, it does not require that the proportionality be expressed by the multiplicative factor $g(\mathbf{x}) = e^{\boldsymbol{\beta}'\mathbf{x}}$; any non-negative function $g(\mathbf{x})$ would do.

However, we could plot against the appropriate function of time to help determine the distribution. For example, if the lifetime distribution is thought to be Weibull, then from $S(t) = \exp\{-(t/\alpha)^\eta\}$, it follows that

$$\ln\{-\ln \hat{S}(t)\} = \eta \ln t - \eta \ln \alpha.$$

The plot of $\ln\{-\ln \hat{S}(t|\mathbf{x})\}$ against $\ln t$ should give a straight line. It is usually easier to see that straight lines are parallel rather than arbitrary curves.

The weakness of this procedure is that the estimates $\hat{S}(t|\mathbf{x})$ will only be satisfactory for this purpose if they are based on sufficiently large numbers of observations; otherwise, the sampling variability will be so large that it might be hard to say whether the curves are parallel or not. This means that there must be a reasonably large number of observations that share the same value of the covariates. For this reason, the method can only be applied if the covariates are few, or by carrying out suitable grouping of values of the covariates.

EXAMPLE 1.1.– *Table 1.1 provides McCool's data on hardened steel specimens tested until failure at four different levels of stress [MCC 80].*

Stress (10^6 psi)	**Ordered lifetimes**
.87 :	1.67 2.20 2.51 3.00 3.90 4.70 7.53 14.70 27.80 37.40
.99 :	0.80 1.00 1.37 2.25 2.95 3.70 6.07 6.65 7.05 7.37
1.09 :	0.012 0.18 0.20 0.24 0.26 0.32 0.32 0.42 0.44 0.88
1.18 :	0.073 0.098 0.117 0.135 0.175 0.262 0.270 0.350 0.386 0.456

Table 1.1. *McCool's data on hardened steel specimens tested until failure at four different levels of stress [MCC 80]*

Figure 1.2 shows the results of carrying out the above graphical procedure on these data. Remember that we are looking for parallel lines describing the sets of points corresponding to these four stress levels in order to confirm the PH assumption. If furthermore they are straight lines, then the Weibull distribution seems to apply. For easier comparison, we have superimposed on the diagram the lines obtained by fitting a Weibull distribution by maximum likelihood to each sample separately. At first sight, it seems very doubtful that the lines are parallel, although one could possibly say that, with only ten observations per group, there will be quite a large sampling variation in the four slopes. However, notice that there is one rather unusual data point, namely the value of 0.012 in the third group, which appears in the bottom left of the diagram. This very early failure seems to be an "outlier", that is, an "observation that is not consistent with the model and the bulk of the data" (see Nelson [NEL 90]). If we omit this point from the fitting, as in Figure 1.3, then the line for the third group is very similar to that of the fourth. All the lines are straight; therefore, it seems that the Weibull model, and hence PH, are reasonable for these data, with the reservation that there is one outlying observation.

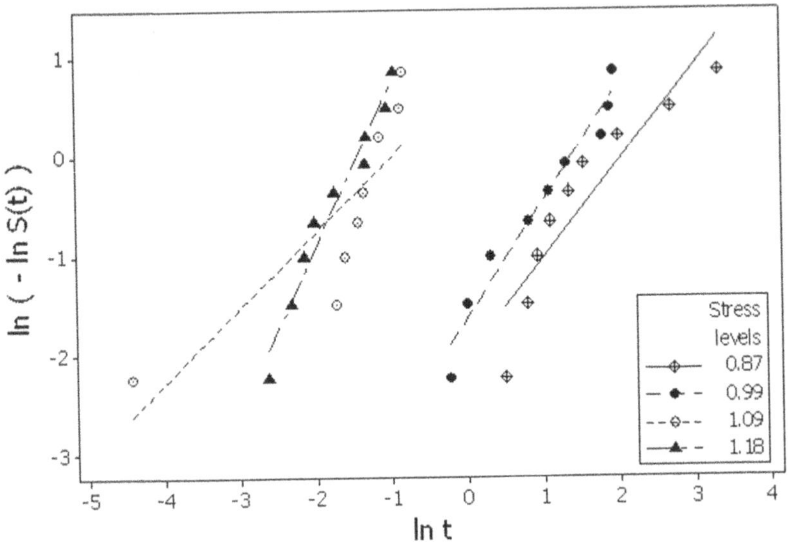

Figure 1.2. *Plot for checking the PH assumption for McCool's data*

We note here that outliers can have a major effect on the fit of statistical models to data, and there is an enormous amount of literature on their detection (see Barnett and Lewis [BAR 94]), although not much of it is applicable to lifetime data. The first step when faced by a possible outlier is to check that the value was recorded correctly and to try to find out if it was recorded under conditions that differed in any way from the rest of the data. If it is confirmed that the value is invalid, then it can be omitted. Otherwise, Nelson suggests that it may be wise to analyze the data with and without this point, to see whether it affects the results appreciably. However, he also points out that "in a sense, suspect data are always right; that is, they reflect something real happening. Only the model or our understanding is inadequate" [NEL 90, p. 209].

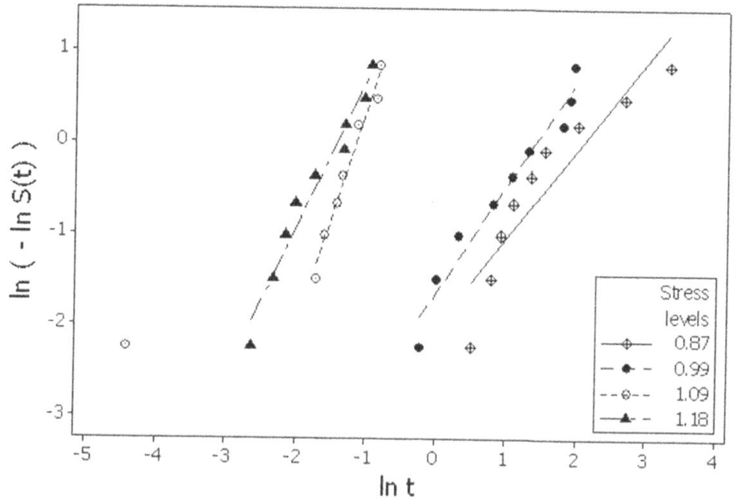

Figure 1.3. *Checking the proportional hazards assumption for McCool's data: one point at bottom left omitted from the fitting*

In section 1.16, we look at another way of checking whether a model provides an adequate description of the data, that is, by examining residuals. As the method is not restricted to PH, but applies equally well to other regression models, we will first look at the other prominent type, accelerated failure time models, and several other less widely used regression models.

1.6. Accelerated failure time models

In order to describe the second main way of introducing dependence on covariates into lifetime models, we begin by considering a model for lifetimes in the form "systematic component + random error" suggested by the general linear regression model given in section 1.3, with the dependent variable the logarithm of lifetime T (thus avoiding the problem of restriction to non-negative values):

$$\ln T_{\mathbf{x}} = \mu + \boldsymbol{\beta}'\mathbf{x} + \sigma\epsilon, \qquad [1.4]$$

where the error term ϵ has location parameter zero and scale parameter 1. The important case of the log-normal distribution arises from taking $\epsilon \sim N(0,1)$ in the regression model for $\ln T_{\mathbf{x}}$. The distribution of $T_{\mathbf{x}}$ is then log-normal for any x. Although the log-normal distribution is widely used because it often provides a good fit to lifetime data, it has a feature that may be unrealistic in many cases, namely that the hazard function increases to a peak and then declines as time increases. On the other hand, the PH property usually implies that the hazard function is monotonically increasing or decreasing (see section 1.4). Thus, we observe that the AFT model incorporates a much wider range of behavior than the PH model is capable of doing. Other distributions for T follow from other assumptions on ϵ. For example, T follows the Weibull distribution if ϵ follows the Gumbel distribution (see section 1.13).

From the above equation for $\ln T_{\mathbf{x}}$,

$$S(t|\mathbf{x}) = P(T_{\mathbf{x}} > t) = P(\mu + \boldsymbol{\beta}'\mathbf{x} + \sigma\epsilon > \ln t), \quad [1.5]$$

hence,

$$\begin{aligned} S(t|\mathbf{x}) &= P\left(\ln T_0 + \boldsymbol{\beta}'\mathbf{x} > \ln t\right) \\ &= P\left(T_0 > te^{-\boldsymbol{\beta}'\mathbf{x}}\right) \\ &= S_0\left(te^{-\boldsymbol{\beta}'\mathbf{x}}\right), \end{aligned}$$

where S_0 is a baseline survival function. Thus, the effect of the covariates $\mathbf{x}$ is to change the time scale. The probability of survival beyond time t, given $\mathbf{x}$, is the same as the baseline probability of survival beyond time $te^{-\boldsymbol{\beta}'\mathbf{x}}$. If $\boldsymbol{\beta}'\mathbf{x} < 0$ this is a longer time than t, and therefore, the survival probability is smaller and the effect of $\mathbf{x}$ is to bring the event forward to shorten lifetimes; hence, the name accelerated failure time (AFT) model. On the other hand, if $\boldsymbol{\beta}'\mathbf{x} > 0$, the effect is to tend to lengthen lifetimes (a deceleration of the time scale).

The last expression above suggests an extended definition of an AFT model in a more general form as $S(t|\mathbf{x}) = S_0(tg(\mathbf{x}))$ for non-negative g, as in section 1.13. This form no longer corresponds to the familiar model [1.4].

Rewriting [1.5]

$$\begin{aligned} S(t|\mathbf{x}) &= P\left(\epsilon > \left(\ln t - \mu - \boldsymbol{\beta}'\mathbf{x}\right)/\sigma\right) \\ &= S_\epsilon\left(\left(\ln t - \mu - \boldsymbol{\beta}'\mathbf{x}\right)/\sigma\right), \end{aligned}$$

where S_ϵ denotes the survival function of an error term ϵ. It follows that

$$f(t|\mathbf{x}) = \frac{1}{\sigma t} f_\epsilon \left(\left(\ln t - \mu - \boldsymbol{\beta}'\mathbf{x} \right) / \sigma \right)$$

and hence the likelihood [1.3] becomes

$$L = \prod_{i=1}^{n} (\sigma t_i)^{-\delta_i} [f_\epsilon(\epsilon_i)]^{\delta_i} [S_\epsilon(\epsilon_i)]^{1-\delta_i}$$

where $\epsilon_i = \left(\ln t_i - \mu - \boldsymbol{\beta}'\mathbf{x}_i \right) / \sigma$.

The AFT formulation has great appeal in the field of reliability, that is, in the engineering and technological applications of lifetime data analysis. In biostatistics, lifetime data are generally obtained from observational studies and only rarely from experimental studies that involve the manipulation of conditions to which units are exposed, with randomized trials providing one notable exception. On the other hand, in reliability, where inanimate objects can be treated in a way that is not possible with human and animal subjects, there is a strong tradition of experimental work. This often involves operating the experimental units under conditions more extreme than will be encountered in normal usage, such as higher temperatures. The purpose is essentially to cause failures to happen quicker than they would be expected to under normal operating conditions. This means that data on a substantial number of failures can be acquired within a rather short space of time. Compare, for example, the lifetimes between the different stress levels in Table 1.1. A more extreme example can be found in Schmee and Hahn's early article about regression with censored data, in which there were no failures at all at the lowest test temperature [SCH 79]. Obtaining many failures at the standard level would either require a study of very long duration - which conflicts with the need to establish results before a product is put on the market or brought into service - or a study including an impractically large number of units.

The design of experiments is, of course, a major field of statistics with a vast literature. Although the general principles of experimental design certainly apply to reliability experimentation, many of the details cannot be carried over easily because of the presence of censoring in reliability data. For extensive material on the design of experiments in reliability, see the books by Meeker and Escobar [MEE 98a] and Nelson [NEL 90].

1.7. Checking the accelerated failure time assumption

A graphical test for the suitability of the accelerated failure time assumption can be derived as follows. The model supposes that

$$S(t|\mathbf{x}) = S_0(tg(\mathbf{x})),$$

where $g(\mathbf{x}) = e^{-\beta'\mathbf{x}}$ in the basic theory and S_0 is the baseline survivor function,

$$\begin{aligned} &= P(T_0 \geq tg(\mathbf{x})) \\ &= P(\ln T_0 \geq \ln t + \ln g(\mathbf{x})) \\ &= S_0^*(y + \ln g(\mathbf{x})), \end{aligned}$$

where $y = \ln t$ and S_0^* is the survivor function of the random variable $Y = \ln T$. This result implies that a plot of $S(t|\mathbf{x})$ against $\ln t$ for particular $\mathbf{x}$ should be a horizontal displacement of S_0^* against $\ln t$. Consequently, the AFT assumption is verified if all the curves $S(t|\mathbf{x})$ for different $\mathbf{x}$ differ from each other only in horizontal displacement when plotted against $\ln t$. To construct these curves, it is necessary to have estimates $\hat{S}(t|\mathbf{x})$, usually Kaplan-Meier. As in the case of the similar graphical test for the PH assumption (see section 1.5), the method is feasible only if there are sufficient data for good estimation of S for each $\mathbf{x}$, or if the $\mathbf{x}$ can be grouped suitably so that each group contains enough observations.

EXAMPLE 1.2.– *Table 1.2 provides Nelson's data on the time to breakdown of an insulating fluid subjected to different voltages [NEL 72]. For simplicity of illustration, three further voltages with a small number of observations are excluded.*

Figure 1.4 shows the plot for checking the AFT assumption for the four groups of observations corresponding to the four voltages. We see that, instead of all the four lines being parallel, the line for the 32 kV level cuts across the others. Hence, the AFT assumption appears to be violated.

The functional form of $g(\mathbf{x})$ could also be investigated graphically. Although each unit's lifetime T follows a different distribution, depending on the value of $\mathbf{x}$, the model requires that the quantities $W = Tg(\mathbf{x})$ all have the same survival function, S_0. Hence, the quantities

$$\ln W = \ln T + \ln g(\mathbf{x})$$

are identically distributed, and consequently, the terms $\ln W$ in the equation

$$\ln T = -\ln g(\mathbf{x}) + \ln W$$

behave like a residual or error term. This means that plotting $\ln t$ against a covariate could indicate the correct functional form of g. For example, if $g(\mathbf{x}) = e^{-\beta'\mathbf{x}}$, then $\ln t$ against x_1 would be a straight line. On the other hand, if x_1 should be replaced by x_1^2, then $\ln t$ against x_1^2 would be a straight line.

Voltage	Failure times					
30kV :	7.74	17.05	20.46	21.02	22.66	43.40
	47.30	139.07	144.12	175.88	194.90	
32kV :	0.27	0.40	0.69	0.79	2.75	3.91
	9.88	13.95	15.93	27.80	53.24	82.85
	89.29	100.58	215.10			
34kV :	0.19	0.78	0.96	1.31	2.78	3.16
	4.15	4.67	4.85	6.50	7.35	8.01
	8.27	12.06	31.75	32.52	33.91	36.71
	72.89					
36kV :	0.35	0.59	0.96	0.99	1.69	1.97
	2.07	2.58	2.71	2.90	3.67	3.99
	5.35	13.77	25.50			

Table 1.2. *Part of Nelson's data on the time to breakdown of an insulating fluid subjected to different voltages [NEL 72]*

Actually, in the experiments in which AFT models are widely used, it is often the case that theory or experience shows the correct functional form for $g(\mathbf{x})$. For example, when the *accelerating factor* (covariate) is temperature, it is common to use the inverse of absolute temperature

$$1/(T + 273.16)$$

or the *Arrhenius transformation*

$$11604.83/(T + 273.16)$$

where the numerator is *Boltzmann's* constant. When the accelerating factor is a load or stress V, then an inverse power relationship $V^{-\alpha}$ is often assumed.

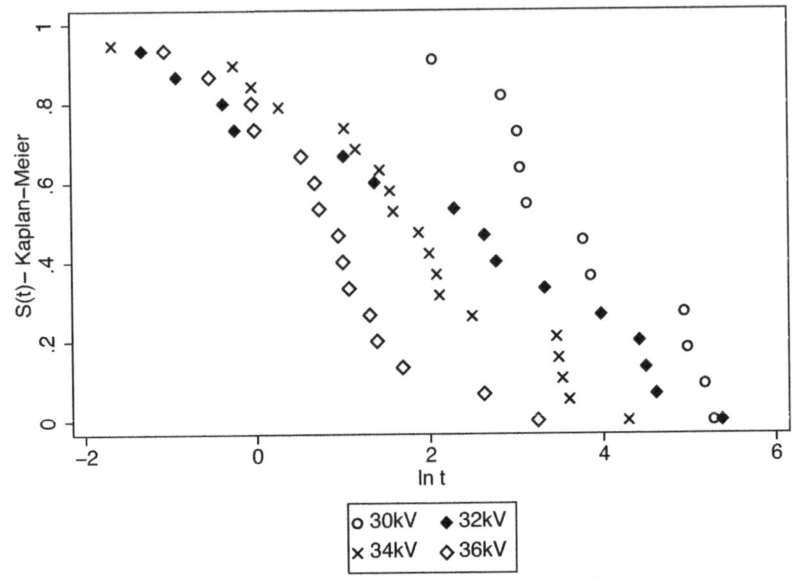

Figure 1.4. *Plot for checking the accelerated failure time assumption in Nelson's data*

1.8. Proportional odds models

Another well-known regression model for lifetime data is the proportional odds (PO) model (see Bennett [BEN 83a, BEN 83b]), which is based on the *odds* of the occurrence of the event by time t:

$$\theta(t) = \frac{F(t)}{1 - F(t)} = \frac{1 - S(t)}{S(t)}.$$

The PO model specifies that

$$\theta(t|\mathbf{x}) = \theta_0(t) g(\mathbf{x}),$$

where $\theta_0(t)$ is the baseline odds and $g(\mathbf{x})$ is a suitable non-negative function as before. The usual choice $g(\mathbf{x}) = e^{\boldsymbol{\beta}'\mathbf{x}}$ gives the model

$$\ln \theta(t|\mathbf{x}) = \ln \theta_0(t) + \boldsymbol{\beta}'\mathbf{x},$$

which is a linear model for log odds, in other words, *logistic regression.* This is the most commonly used regression model for binary data. Despite this appealing link, PO models have been used infrequently compared to others since they were introduced by Bennett [BEN 83a, BEN 83b]. With the exception of the regression model based on the log-logistic distribution, which is both a PO model and an AFT model (see below), they are mathematically and computationally more difficult to handle. Thus, they have not been included in some of the computing packages, and therefore, we will not be discussing them in detail here. Detailed discussions of PO models can be found in Collett [COL 14] and Dauxois and Kirmani [DAU 03], for example. The latter presents a graphical procedure for examining the hypothesis of PO between groups of respondents and a formal test for the case of two groups.

In order to see an important difference between PO and PH models, consider the hazard function of a PO model. Solving the previous equations for $S(t|\mathbf{x})$ gives

$$S(t|\mathbf{x}) = \{1 + \theta_0(t)g(\mathbf{x})\}^{-1}$$

from which the ratio of the hazard functions for two units with different covariate values is

$$\frac{h(t|\mathbf{x}_1)}{h(t|\mathbf{x}_2)} = \frac{g(\mathbf{x}_1)}{1 + \theta_0(t)g(\mathbf{x}_1)} \Big/ \frac{g(\mathbf{x}_2)}{1 + \theta_0(t)g(\mathbf{x}_2)}$$

$$\to 1 \quad (t \to \infty)$$

because $\theta_0(t) \to \infty$ as $t \to \infty$ (since $S_0(t)$ tends to zero), whereas $g(.)$ does not change with t. This fact - that the hazards for two different units tend to equalize over time under the PO model - stands in major contrast to the PH property that two units' hazards remain in the same ratio forever. On the one hand, the PO model says that initial differences disappear; on the other hand, the PH model says that they never change. Depending on the context, either postulate might be more appropriate. A badly made unit remains badly made; therefore, the implication of the PO model seems unreasonable in such situations. However, some treatments that a patient receives may only have a temporary effect that wears off with time, and in a case like that, it is the PH property - that initial differences in the covariates continue to have the same effect for ever and ever - that seems unrealistic.

All distributions of the Marshall-Olkin extended form [MAR 97] have the PO property (see also [ECO 07, SAN 08, CAR 10]). The Marshall-Olkin formula is

$$S^*(t|\alpha) = \frac{\alpha S(t)}{1 - \bar{\alpha} S(t)}, \qquad [1.6]$$

where S is the survival function of the original distribution and S^* is the survival function of the new, extended distribution; $\alpha > 0$ is a constant, and $\bar{\alpha} = 1 - \alpha$. Applying the formula again to [1.6] just results in a different value of α; therefore, the family of distributions is closed. However, the PO model states that

$$\frac{1 - S(t|\mathbf{x})}{S(t|\mathbf{x})} = g(\mathbf{x}) \frac{1 - S_0(t)}{S_0(t)},$$

which gives

$$S(t|\mathbf{x}) = \frac{\frac{1}{g(\mathbf{x})} S_0(t)}{1 - \left(1 - \frac{1}{g(\mathbf{x})}\right) S_0(t)},$$

and this is just the Marshall-Olkin form with $\alpha = 1/g(\mathbf{x})$. Therefore, all Marshall-Olkin extended distributions are PO distributions. For example, the Marshall-Olkin extended Weibull distribution (see Ghitany *et al.* [GHI 05]; Caroni [CAR 10]) is a PO distribution, although the Weibull distribution itself is not a PO distribution.

Zucker and Yang [ZUC 06] note that the PO and PH models are both special cases of the general form

$$h(S(t|\mathbf{x})) = h(S_0(t)) \exp(\boldsymbol{\beta}'\mathbf{x}),$$

where h is a suitable monotonically decreasing function from $[0,1]$ to $[0,\infty]$. For the PH model, $h(s) = -\ln(s)$, and for the PO model, $h(s) = (1-s)/s$. These are both included within the Box-Cox family

$$h(s,\rho) = \rho^{-1}\left(s^{-\rho} - 1\right),$$

with the PH model arising when $\rho \to 0$ and the PO model when $\rho = 1$. Zucker and Yang give references to previous appearances of this family in the survival analysis literature; their contribution consists of two estimation methods, although they observe that the model essentially cannot be fitted if the event rate is low.

Further work on the PO model is reviewed by Chen *et al.* [CHE 12], who extend the model by incorporating external time-varying covariates (see section 4.2).

1.9. Proportional mean residual life models

The mean residual life (MRL) of a unit is defined conditionally on the unit's present age t and represents its expected lifetime beyond this point,

$$\mu(t) = E[T - t|T \geq t],$$

which, if it exists, is equal to

$$\mu(t) = \frac{\int_t^\infty S(u)du}{S(t)},$$

as in section 1.1. The proportional mean residual life (PMRL) model proposes the relationship

$$\mu(t|\mathbf{x}) = g(\mathbf{x})\mu_0(t),$$

where $\mu_0(t)$ is the baseline MRL function.

Oakes and Dasu [OAK 90, OAK 03] suggest that the MRL function provides a more natural basis for modeling lifetime data than the hazard function because it summarizes the entire remaining life distribution and not just the immediate risk of failure. They claim that this is likely to be the more important information for the design of maintenance and repair strategies. The MRL is also used extensively in demography under the name life expectancy. On the other hand, Hougaard [HOU 99] states that, in contrast to these industrial and demographic applications, the evaluation of mean lifetime in biostatistics is considered unacceptable. He gives three reasons for this: one, the difficulty of estimating the right tail of the residual lifetime distribution (exacerbated by censoring), which - as acknowledged by Oakes and Dasu [OAK 03] - can have a strong influence on the mean; two, the possible tendency for readers to think in terms of the normal distribution when they are presented with means, which could be very misleading; and three, the fact that for some types of events, there may be a proportion of the population that will never experience the event (see section 1.4 and elsewhere), which makes it impossible to calculate a mean. The last of these objections obviously does not apply in industrial and demographic applications, because units will always fail and individuals will always die eventually.

Further development of this model was taken up by Maguluri and Zhang [MAG 94] and subsequently by others; see Chen and Cheng [CHE 05]. The PMRL model does not seem to have entered general use at present. It is noticeable that an extensive review of statistical estimation of the remaining useful life of an item (Si *et al.* [SI 11]) does not mention the PMRL model at all.

1.10. Proportional reversed hazard rate models

Yet another "proportional" model is the proportional reversed hazard rate (PRHR) model (see Gupta and Gupta [GUP 07]), which is defined in a closely similar way to the PH model. The reversed hazard rate $r(t)$ is related to the conditional probability that an event occurred in the interval of length δt *before* time t, in contrast to the hazard that is related to the occurrence of the event in the interval of length δt *after* time t as in equation [1.1]. Thus,

$$r(t)\delta t = P(t - \delta t < T \leq t | T \leq t) = f(t)\delta t / F(t),$$

and therefore, $r(t) = (d/dt) \ln F(t)$. The PRHR model defines a multiplicative effect of covariates on the baseline function $r_0(t)$,

$$r(t|\mathbf{x}) = g(\mathbf{x}) r_0(t).$$

Equivalently, the model may also be defined by the relation $F(t|\mathbf{x}) = [F_0(t)]^{g(\mathbf{x})}$ similar to the definition $S(t|\mathbf{x}) = [S_0(t)]^{g(\mathbf{x})}$, which can be derived for the PH model.

An example of a family of distributions with the PRHR property is the exponentiated Weibull, which has baseline distribution function $F(t|\alpha) = [1 - \exp(-t^{\alpha})]^{\theta}$ [MUD 96].

A PRHR model was first suggested, although not studied, by Kalbfleisch and Lawless [KAL 89], who examined a problem in which occurrences were ascertained after the event, in which case the retrospective nature of the reversed hazard rate has a natural appeal.

1.11. The accelerated hazards model

The accelerated hazards (AccH) model incorporates ideas from both the PH and AFT models. As seen earlier, the effect of covariates in the PH model can be simply expressed by a multiplicative effect on the hazard function: in its usual form, the model is

$$h(t|\mathbf{x}) = h_0(t)e^{\boldsymbol{\beta}'\mathbf{x}}.$$

In the AFT model, the effect of covariates is a shift of time scale, so that the survival function becomes

$$S(t|\mathbf{x}) = S_0\left(te^{\boldsymbol{\beta}'\mathbf{x}}\right).$$

Adapting these ideas, the AccH model takes the time scale shift and places it in the hazard function

$$h(t|\mathbf{x}) = h_0\left(te^{\boldsymbol{\beta}'\mathbf{x}}\right).$$

This model was introduced by Chen and Wang [CHE 00] for a two-group comparison and extended to the more general regression case by Chen [CHE 01a]. The motivation for the model was provided by a clinical trial, in which there appeared to be no immediate difference in hazard rates between the two treatment groups, whereas both the PH and AFT models imply that covariates have an effect even at time zero. This implication is somewhat problematic for randomized clinical trials, in which the groups should not differ at baseline and the treatment is unlikely to have an instant effect. As the trial progressed, an increasing difference between hazard rates was seen, which is again in conflict with the PH property. Furthermore, after some time, the two hazard functions crossed over, which is a feature that the PH model cannot reproduce, although the AFT model can; neither PH nor AFT allows survival functions to cross. The new AccH model, on the other hand, does allow hazard functions and survival functions to cross over, depending on the form of the baseline hazard function. Conditions for crossovers in hazard functions in the various models are given by Zhang and Peng [ZHA 09].

In the AFT model, the hazard function is

$$h(t|\mathbf{x}) = e^{\boldsymbol{\beta}'\mathbf{x}} h_0 \left(t e^{\boldsymbol{\beta}'\mathbf{x}} \right),$$

which suggests a generalized model encompassing all the three PH, AFT and AccH models:

$$h(t|\mathbf{x}) = h_0 \left(t e^{\boldsymbol{\beta}_1'\mathbf{x}} \right) e^{\boldsymbol{\beta}_2'\mathbf{x}}.$$

The PH model corresponds to $\boldsymbol{\beta}_1 = \mathbf{0}$, the AFT model to $\boldsymbol{\beta}_1 = \boldsymbol{\beta}_2$ and the AccH model to $\boldsymbol{\beta}_2 = \mathbf{0}$. This model and its semi-parametric estimation is studied by Chen and Jewell [CHE 01b], but in fact, the model had been introduced and analyzed several years earlier by Ciampi and Etezadi-Amoli [CIA 85, ETE 87] under the name extended hazard regression. Subsequently, Shyur *et al.* discussed its applicability in the field of reliability [SHY 99].

The two vectors of regression coefficients can be interpreted as measuring different impacts of the covariates on survival. While the appropriate component of $\boldsymbol{\beta}_1$ measures a specific covariate's contribution to the acceleration (or deceleration) factor, the corresponding component of $\boldsymbol{\beta}_2$ indicates its independent contribution to the relative hazards. In the example used for illustration of the method by Etezadi-Amol and Ciampi (survival of patients with ovarian cancer), two of the five covariates appeared to have a simple PH effect; one had an AFT effect, and the remaining two had both the effects. One of these, the patient's age, had opposite signs of its two coefficients, suggesting an increased hazard for older patients at any given age, but slower tumor growth in older patients than in younger ones. This structure, in which a covariate can affect lifetimes in two ways, has some similarity to the first hitting time regression model based on an underlying Wiener process, which is the main topic of this book and will be introduced in the next chapter. The issue of opposing effects indicated by the signs of the two coefficients associated with the same covariate will be mentioned in that context too.

Although the general model may give a better description of the data than any of the three separate models included in it, Chen and Jewell [CHE 01b] suggest that its main value may lie in bringing out the differences between these three models. Earlier, Chen [CHE 01a] commented on the potential usefulness of fitting the general model as a guide to which of the separate models to fit. Of course, this will only become true in practice when the readily available software makes it easy to fit the more general model, which is not the case at the moment.

1.12. The additive hazards model

It can often be supposed that an organism or machine comprises many components or parts that must *all* be operating; otherwise the organism dies or the machine fails (a series system). If a particular component j has hazard rate $h_j(t)$ at time t, then the overall hazard $h(t)$ at this time is $h(t) = \sum_j h_j(t)$. Consequently, it may often seem natural to represent a hazard function in additive form (see Elandt-Johnson [ELA 80]).

In the additive hazards model, the effect of the covariates on the baseline hazard function $h_0(t)$ is additive

$$h(t|\mathbf{x}) = h_0(t) + \boldsymbol{\beta}'\mathbf{x}$$

instead of multiplicative as in the PH model. This model was first suggested by Aalen [AAL 78] and further developed by Aalen [AAL 89] and others. Lin and Ying [LIN 94] proposed a semi-parametric estimation method with $h_0(t)$ unspecified, along the lines of the semi-parametric PH model. They take the regression parameters $\boldsymbol{\beta}$ as fixed, whereas Aalen's formulation allowed for time-varying coefficients $\boldsymbol{\beta}(t)$.

For an example of the application of the additive hazards model in a biostatistical context, with comparison of results between its different versions and also with the Cox model, see Xie *et al.* [XIE 13]. They point out that the multiplicative and additive hazards model address different questions. While the PH model provides estimates of relative hazard, the additive hazards model estimates absolute differences in hazard. Therefore, in the latter case and assuming that the event rate is low, the differences between cumulative hazards give an approximation to differences in cumulative incidence. Thus, an estimate of attributable risk is obtained. This information could be important for the purpose of public health planning and intervention.

The application of the additive hazards model in the reliability context, especially concerning repairable systems, is considered by Pijnenburg [PIJ 91].

1.13. PH, AFT and PO distributions

Based on our choice of how to introduce covariates (PH, AFT or PO - we will not be considering other models any further), we will find that some lifetime distributions are much easier to use than others. Let us begin with the Weibull distribution in a PH model. Since the baseline hazard is

$$h_0(t) = \frac{\eta t^{\eta-1}}{\alpha^\eta}, \ t > 0,$$

we have

$$h(t|\mathbf{x}) = h_0(t)g(\mathbf{x}) = \eta t^{\eta-1}\alpha^{-\eta}e^{\boldsymbol{\beta}'\mathbf{x}} = \frac{\eta t^{\eta-1}}{\left(\alpha e^{-\boldsymbol{\beta}'\mathbf{x}/\eta}\right)^\eta},$$

which corresponds to another Weibull distribution. The shape parameter η is the same as that of the baseline hazard, but the scale parameter is different: α has become $\alpha e^{-\boldsymbol{\beta}'\mathbf{x}/\eta}$. Thus, PH implies that the lifetime distribution for any unit is always Weibull when the baseline hazard is Weibull, and in this sense, the Weibull is a "PH distribution", and is therefore a natural choice to use in the context of a PH model.

The AFT model in section 1.6 was formulated as a model for the logarithm of T. If $T \sim Weibull(\alpha, \eta)$, then $\ln T \sim Gumbel$ with

$$S(t) = \exp\left(-e^{(\ln t - \mu)/\sigma}\right),$$

where $\mu = \ln \alpha$, $\sigma = \eta^{-1}$. This implies that, if $\epsilon \sim Gumbel(0, 1)$ in the AFT model

$$\ln T_{\mathbf{x}} = \mu + \boldsymbol{\beta}'\mathbf{x} + \sigma\epsilon,$$

then $T_{\mathbf{x}} \sim Weibull$ for any $\mathbf{x}$. Therefore, the Weibull is also an "AFT distribution", in addition to being a PH distribution; in fact, no other distribution has this dual property (see below). The Weibull distribution is not a PO distribution. As noted by Hougaard, the PH model modifies the

Gompertz distribution with parameters μ and ϕ into another Gompertz distribution with the same ϕ but different μ. However, the AFT model changes both the parameters [HOU 99]. Thus, the PH and AFT models are not equivalent for the Gompertz distribution.

Other important AFT distributions, which are not also PH, include the log-normal and log-logistic distributions. T has a log-logistic distribution when $Y = \ln T$ follows a logistic distribution. The survival function of the log-logistic distribution is

$$S(t) = (1 + e^{\kappa} t^{\gamma})^{-1}, \ \gamma > 0, \ \kappa > 0,$$

and the hazard function is

$$h(t) = \frac{\gamma t^{\gamma-1} e^{\kappa}}{1 + e^{\kappa} t^{\gamma}}.$$

We take this as the baseline hazard function $h_0(t)$ in the following. Now for any AFT model, we have

$$\begin{aligned} h(t|\mathbf{x}) = \frac{f(t|\mathbf{x})}{S(t|\mathbf{x})} &= \frac{-d \ln S(t|\mathbf{x})}{dt} \\ &= \frac{-d \ln S_0(te^{-\boldsymbol{\beta}'\mathbf{x}})}{dt} \\ &= e^{-\boldsymbol{\beta}'\mathbf{x}} h_0(te^{-\boldsymbol{\beta}'\mathbf{x}}); \end{aligned}$$

therefore, in the case of the log-logistic distribution, we obtain

$$\begin{aligned} h(t|\mathbf{x}) &= e^{-\boldsymbol{\beta}'\mathbf{x}} h_0(te^{-\boldsymbol{\beta}'\mathbf{x}}) \\ &= \frac{e^{-\boldsymbol{\beta}'\mathbf{x}} \gamma (te^{-\boldsymbol{\beta}'\mathbf{x}})^{\gamma-1} e^{\kappa}}{1 + e^{\kappa} (te^{-\boldsymbol{\beta}'\mathbf{x}})^{\gamma}} \\ &= \frac{\gamma t^{\gamma-1} e^{\kappa - \gamma \boldsymbol{\beta}'\mathbf{x}}}{1 + (e^{\kappa - \gamma \boldsymbol{\beta}'\mathbf{x}}) t^{\gamma}}, \end{aligned}$$

which is still the hazard function of a log-logistic distribution, although the parameter κ has changed to $\kappa - \gamma \boldsymbol{\beta}'\mathbf{x}$, with the shape parameter γ remaining unchanged. Hence, the log-logistic is an AFT distribution.

In the case of the PO model, a PO distribution will be one for which the distribution of

$$\theta(t|\mathbf{x}) = g(\mathbf{x})\theta_0(t)$$

has the same functional form as the baseline odds function $\theta_0(t)$. The simplest example is the log-logistic distribution with survival function given above and odds $e^{\kappa}t^{\gamma}$. Then

$$\theta(t|\mathbf{x}) = g(\mathbf{x})e^{\kappa}t^{\gamma} = e^{\kappa+\ln g(\mathbf{x})}t^{\gamma},$$

which corresponds to a log-logistic distribution with parameter κ changed to $\kappa + \ln g(\mathbf{x})$ and shape γ unchanged. Thus, the log-logistic distribution is both PO and AFT; in fact, it is unique in this respect (see below). It is not, however, PH.

Proof of the unique status of the Weibull and log-logistic distributions in possessing dual properties (PH and AFT for the Weibull, PO and AFT for the log-logistic) is given by [LAW 86]. Define a general family of regression models by

$$\psi_1\left\{S_{\mathbf{x}}(t)\right\} = \psi_1\left\{S_0(t)\right\} + g_1(\mathbf{x}),\ t > 0,$$

where $S_{\mathbf{x}}(t)$ is a survival function for a unit with vector of covariates $\mathbf{x}$ and baseline $S_0(t)$, and $g_1(\mathbf{0}) = 0$. This family includes PH models, for which

$$\psi_1(u) = \ln\left(-\ln u\right) \qquad [1.7]$$

and also PO models, choosing

$$\psi_1(u) = \ln\left\{(1-u)/u\right\}. \qquad [1.8]$$

Furthermore, define a second family by

$$\psi_2\left\{Q_{\mathbf{x}}(p)\right\} = \psi_2\left\{Q_0(p)\right\} + g_2(\mathbf{x}),\ \ 0 < p < 1,$$

where $Q_{\mathbf{x}}(p)$ is the quantile function of T given $\mathbf{x}$, with baseline $Q_0(p)$, and $g_2(\mathbf{0}) = 0$. This family includes AFT models, for which $\psi_2(u) = \ln u$.

Lawless [LAW 86] shows that for given ψ_1 and ψ_2, the unique family satisfying both [1.7] and [1.8] simultaneously is

$$S_0(t) = \psi_1^{-1}\left\{a\psi_2(t) + b\right\}.$$

In particular, choosing $\psi_1(u) = \ln(-\ln u)$ (a PH model) and $\psi_2(u) = \ln u$ (an AFT model) gives

$$S_0(t) = \exp\left(-e^b t^a\right),$$

where $a \neq 0$ and b are constants, which is a Weibull model. Hence, the Weibull distribution, and no other, is both PH and AFT. The same ψ_2 and $\psi_1(u) = \ln\left\{(1-u)/u\right\}$ (a PO model) gives

$$S_0(t) = \left(1 + e^b t^a\right)^{-1},$$

which is a log-logistic model. Hence, the log-logistic distribution, and no other, is both AFT and PO.

1.14. Cox's semi-parametric PH regression model

We present in this section the widely used version of the PH regression model known as Cox's semi-parametric regression model. First, we recall from section 1.4 that the PH property requires that the hazard functions of two units, with covariate vectors $\mathbf{x}_1$ and $\mathbf{x}_2$, must be in constant ratio to each other over time. This is achieved if the hazard function takes the form

$$h(t|\mathbf{x}) = h_0(t)g(\mathbf{x}),$$

where $h_0(t)$ is a baseline hazard and $g(\mathbf{x})$ is a suitable non-negative function of $\mathbf{x}$. The baseline could be, for example, the hazard that applies to a unit with $\mathbf{x} = \mathbf{0}$ (although in many applications, this value cannot be realized). Cox [COX 72] proposed taking the form of $g(\mathbf{x})$ that we have already seen

$$g(\mathbf{x}) = e^{\boldsymbol{\beta}'\mathbf{x}}$$

but - and this is the model's crucial feature - not specifying the functional form of h_0 at all. Only the part of the model that expresses dependence on the covariates is expressed parametrically (hence the description of the model as semi-parametric) and in fact, only this part needs to be estimated.

Consider the set of units R_i that are at risk of failure at time $t_{(i)}$, where $t_{(1)} < t_{(2)} < \cdots < t_{(k)}$ are unique failure times recorded in a study of $n \geq k$ units. This risk set R_i excludes any units that have already failed or were censored earlier than $t_{(i)}$. From the definition of hazard rate, the probability that unit $j \in R_i$ with covariates $\mathbf{x}_j$ fails in the time interval $(t_{(i)}, t_{(i)} + \delta t)$ is

$$h_0(t_{(i)})\, e^{\boldsymbol{\beta}'\mathbf{x}_j}\, \delta t.$$

Hence, the conditional probability that it is unit j that fails at time $t_{(i)}$, given that we know that one unit does fail at this time, is

$$\frac{h_0(t_{(i)})\, e^{\boldsymbol{\beta}'\mathbf{x}_j}\, \delta t}{\sum_{\ell \in R_i} h_0(t_{(i)})\, e^{\boldsymbol{\beta}'\mathbf{x}_\ell}\, \delta t} = \frac{e^{\boldsymbol{\beta}'\mathbf{x}_j}}{\sum_{\ell \in R_i} e^{\boldsymbol{\beta}'\mathbf{x}_\ell}},$$

in which h_0 ultimately does not appear at all. Taking the product of these conditional probabilities over all the failure times $\{t_{(i)} : i = 1, \ldots, k\}$ gives the partial likelihood

$$\prod_{i=1}^{k} \left\{ \frac{e^{\boldsymbol{\beta}'\mathbf{x}_i}}{\sum_{\ell \in R_i} e^{\boldsymbol{\beta}'\mathbf{x}_\ell}} \right\}.$$

This is treated as a standard likelihood, which is maximized over $\boldsymbol{\beta}$ in order to obtain estimates of the effects of the covariates on the hazard rate, without needing to estimate the hazard rate itself (although a non-parametric estimate of $h_0(t)$ can be obtained if desired). The theory, which can be extended to allow for tied failure times, is presented in detail in Therneau and Grambsch [THE 00] and elsewhere. The justification for treating the partial likelihood as a standard likelihood was presented intuitively in Cox's initial paper on this regression method [COX 72] and was subsequently placed on a firmer foundation by Cox [COX 75] and others.

The remarkable simplicity that is made possible by using the partial likelihood method is one of the features behind the widespread adoption of Cox's semi-parametric PH regression model (often referred to simply as Cox regression), at least in the biomedical sciences where it has become virtually the default approach toward analyzing lifetime data. It also has an advantage of apparently easy interpretability. From $h(t|\mathbf{x}) = h_0(t)e^{\boldsymbol{\beta}'\mathbf{x}}$, the effect of a unit increase in the value of the jth covariate x_j (the familiar way of interpreting the size of a regression coefficient) is to multiply the hazard by $e^{\hat{\beta}_j}$. This is very similar to the interpretation of the coefficients of logistic regression, which is another almost-default method of analysis of biomedical data.

In sharp contrast to the remarkably high level of use of Cox's PH model in biostatistics, it is not often seen in the field of reliability. A few illustrative applications can be found in, for example, Bendell *et al.* [BEN 86, BEN 91], Dale [DAL 85], Elsayed and Chan [ELS 90], Krivtsov *et al.* [KRI 02] and Madeira *et al.* [MAD 13]. Some reviews warn against the unthinking transfer of biostatistical methods to reliability analysis (see Bendell *et al.* [BEN 91] and Kumar and Klefsjö [KUM 94]), particularly because of the preponderance of repairable systems in the latter field, and hence the need to find realistic models for recurrent events, that is, repeated events occurring in the same unit. The preference in the field of reliability for parametric AFT models is undoubtedly related to the importance of analysis of experiments in which one or more factors (such as the operating temperature) have been manipulated specifically in order to bring forward the times of failure. The concept of the AFT formulation is a natural fit to the nature of such data. Elsayed and Chan remark in their presentation of Cox modeling in a reliability problem in electronics that "The proportional hazards relation has not been used much for modeling the hazard (failure) rate of electronic devices because there appears to be no physical basis for hazard-rate scaling" [ELS 90, p.331], in contrast to the familiar concept of time scaling in the AFT model. However, they go on to interpret the PH relation in terms of one of the reliability models taken from the literature.

1.15. PH versus AFT

The dominance of PH models in biostatistics and AFT in reliability has led to much consideration of the differences between the two and recommendations for which model is preferable in the given circumstances. The rapid adoption of Cox's PH model in biostatistics to the relative neglect of the AFT model, and the converse situation in reliability, gave rise to a number of articles in the literature, whose main purpose is to encourage readers who are used to one formulation not to ignore the other. The papers in the reliability literature on the Cox model - for example, those cited in section 1.14 - tend to be of this type. Conversely, papers appear in the biostatistical literature promoting the virtues of alternatives to the Cox model, notably AFT models. One example where an AFT model makes more sense than a PH model is in a trial of a drug for influenza (see Kay and Kinnersley [KAY 02] and Patel *et al.* [PAT 06]). The effect of the drug is to shorten the duration of the illness in treated patients compared to untreated patients, but almost all the patients would be expected to recover from their symptoms during the course of the trial, irrespective of the treatment. Therefore, a PH model - with constant hazards throughout - is an illogical structure, whereas the AFT framework describes the effect (acceleration of recovery) more appropriately. In another example, Argyropoulos *et al.* [ARG 09] discuss the survival of hemodialysis patients and argue against the PH model because it evaluates the effect of a covariate at particular time points rather than considering its history. If a covariate acts through "accumulated damage", then AFT rather than PH would be appropriate.

In situations where prior considerations lead to preference for neither model over the other, model-checking methods should be applied as in all the applications of statistical analyses. A wide range of diagnostics - some based on analogies with linear regression, others founded on the particular properties of the survival analysis context - exists for the Cox PH model (see, for example, Therneau and Grambsch [THE 00] and Caroni [CAR 04]). Nardi and Schemper [NAR 03] illustrate the use of residuals in examining Cox and parametric AFT models. However, most published applications do not present an investigation of the validity of the assumptions underlying Cox's model. For example, Altman *et al.* [ALT 95] found that the assumptions were checked in only 2/43 (5%) of the papers that they examined, and none assessed goodness of fit. Similarly, Ford *et al.* [FOR 95, p.745] stated that "model validation is an important prerequisite to the interpretation of parameter estimates. In this respect, the almost de facto assumption of the Cox model in the analysis of survival data is a cause for concern".

If the basic PH assumption appears to be violated, a number of solutions are available while remaining within a general PH framework (see Therneau and Grambsch [THE 00]). One popular method is stratification. If the PH assumption does not hold for a particular covariate, then the model can be fitted (without any additional technical difficulty) stratifying by values of that covariate. The model in the mth stratum becomes

$$h_m(t|\mathbf{x}) = e^{\boldsymbol{\beta}'\mathbf{x}} h_{m0}(t),$$

where $h_{m0}(t)$ is the baseline hazard function in stratum m, thus allowing different baseline hazards in each stratum, but assuming common effects of the other covariates in every stratum. With obvious extension of the notation of section 1.14, the log partial likelihood for events in the mth stratum is

$$\ell_m(\boldsymbol{\beta}) = \sum_{i=1}^{k_m} \boldsymbol{\beta}'\mathbf{x}_{mi} - \sum_{i=1}^{k_m} \ln \left\{ \sum_{\ell \in R_{mi}} e^{\boldsymbol{\beta}'\mathbf{x}_{m\ell}} \right\},$$

and the overall log partial likelihood for all s strata is

$$\ell(\boldsymbol{\beta}) = \sum_{m=1}^{s} \ell_m(\boldsymbol{\beta}) = \sum_{m=1}^{s} \sum_{i=1}^{k_m} \boldsymbol{\beta}'\mathbf{x}_{mi} - \sum_{m=1}^{s} \sum_{i=1}^{k_m} \ln \left\{ \sum_{\ell \in R_{mi}} e^{\boldsymbol{\beta}'\mathbf{x}_{m\ell}} \right\}.$$

This can be maximized with no more difficulty than for the single-stratum model. A drawback of this formulation is that it does not yield a direct estimate of the effect of the stratification factor on lifetimes.

Important general results on the properties of PH and AFT models have been obtained, particularly with regard to the robustness of estimates against misspecification of the model. It is known that omitting a relevant covariate from a model that is truly PH induces a model that is no longer PH. The estimated regression coefficients, their standard errors and the ratios of coefficients to standard errors all tend to be smaller than the corresponding quantities obtained by estimation under the true model. Unlike linear regression, this is true even if the omitted covariates are orthogonal to those that are included. References relevant to this topic include Gail *et al.* [GAI 84], Struthers and Kalbfleisch [STR 86], Schumacher *et al.* [SCH 87], Schmoor and Schumacher [SCH 97] and Gerds and Schumacher [GER 01]. An extensive study in the context of fully parametric PH and AFT models was reported by Hutton and Monaghan [HUT 02]. A key conclusion is that estimates from a misspecified PH model can be seriously biased, and the apparent shape of the hazard function can be misleading. Furthermore, the size of Wald tests is underestimated. On the other hand, AFT models are more robust to misspecification, a property which is attributed to their log-linear form. Hougaard [HOU 99, p. 22] says "the accelerated failure parameter η is robust toward neglected covariates, whereas the proportional hazards parameter β is not .... It is a major drawback of the PH model...". As far as estimated survival is concerned, bias in the lower and upper percentiles can be substantial from a misspecified model, less so for the median.

There are however results that show that some similarity of results between PH and AFT regressions can be expected. Fitting a PH model when the true model is AFT leaves the relative importance of covariates unchanged to first order under conditions (see Solomon [SOL 84] and Struthers and Kalbfleisch [STR 86]). Under the AFT model, the hazard function given covariates $\mathbf{x}$ takes the form

$$h_{AFT}(t|\mathbf{x}) = e^{\boldsymbol{\beta}'\mathbf{x}} h_0(e^{\boldsymbol{\beta}'\mathbf{x}} t)$$

for baseline hazard h_0. Now, following Kwong and Hutton, take a Taylor series expansion in $e^{\boldsymbol{\beta}'\mathbf{x}}$ about $\boldsymbol{\beta} = \mathbf{0}$ to first order:

$$\begin{aligned} h_{AFT}(t|\mathbf{x}) &\approx e^{\boldsymbol{\beta}'\mathbf{x}} h_0 \left\{(1 + \boldsymbol{\beta}'\mathbf{x})\, t\right\} \\ &= e^{\boldsymbol{\beta}'\mathbf{x}} h_0(t + t\,\boldsymbol{\beta}'\mathbf{x}) \\ &\approx e^{\boldsymbol{\beta}'\mathbf{x}} \left\{h_0(t) + \boldsymbol{\beta}'\mathbf{x}\, t\, h_0'(t)\right\}, \end{aligned}$$

where the second approximation is obtained from a first order Taylor series expansion about t. Consequently, we have

$$\begin{aligned} h_{AFT}(t|\mathbf{x}) &\approx e^{\boldsymbol{\beta}'\mathbf{x}} h_0(t) + e^{\boldsymbol{\beta}'\mathbf{x}} \boldsymbol{\beta}'\mathbf{x}\, t\, h_0'(t) \\ &= h_{PH}(t|\mathbf{x}) + e^{\boldsymbol{\beta}'\mathbf{x}} \boldsymbol{\beta}'\mathbf{x}\, t\, h_0'(t), \end{aligned}$$

where $h_{PH}(t|\mathbf{x})$ is the hazard function that holds for a unit with covariates $\mathbf{x}$ under the PH assumption. This expression implies that hazards derived under AFT and PH will not differ greatly so long as (a) covariate effects $\boldsymbol{\beta}$ are small, and (b) the hazard function h_0 varies slowly so that $t\, h_0'(t)$ is small [KWO 03].

As the Cox model is semi-parametric, it might be expected that it would yield less efficient parameter estimates than an appropriate fully parametric model. This has been investigated by Oakes [OAK 77] and Efron [EFR 77], among others. Nardi and Schemper [NAR 03] summarize the conditions for this to be true as follows:

a) if parameter values are not close to zero;

b) if follow-up depends on the values of the covariates;

c) if the covariates show a strong time trend.

Furthermore, as expected, the loss in precision is greater for small samples.

It is often remarked that the PH model is not based on any persuasive rationale, and that its popularity is to a large extent due to its apparent simplicity. For example, according to Oakes [OAK 13, p.453]: "Cox (1972) emphasized that there is usually no simple physical or biological motivation for the assumption of PH. The appeal of this model arises rather from the intuitive interpretation of the hazard ratio in terms of conditional risks, and from the simplicity and numerical stability of the algorithms used to fit the model. However in particular situations other approaches may be preferable". Cox himself, in his original presentation, claimed that his model was "intended as a representation of the behaviour of failure-time that is convenient, flexible and yet entirely empirical" [COX 72, p.200], and concluded the paper with the claim that the model "as a basis for rather empirical data reduction ... seems flexible and satisfactory" [COX 72, p.201]. However, Freedman's objection seems entirely reasonable: "if the model is wrong, why are the parameter estimates a good summary of the data?" [FRE 08, p.117].

1.16. Residuals

One basic way of checking the suitability of a statistical model is to inspect the *residuals* after the model has been fitted. Examination of the residuals can show if the model's assumptions are satisfied and how well the model fits the data, not just overall but for each point individually. In the familiar case of linear regression, the residuals are

$$\begin{aligned} \hat{\epsilon}_i &= y_i - \hat{y}_i \\ &= y_i - \hat{\boldsymbol{\beta}}' \mathbf{x}_i, \end{aligned}$$

in other words, the difference between the observed value y_i and the predicted $\hat{y}_i$. These residuals can be examined in various ways, often graphically. For example, their distribution can be investigated, or possible outliers can be identified.

However, most statistical models do not give rise to residuals of this familiar form. Consequently, Cox and Snell [COX 68] proposed *generalized residuals*. Suppose that the random variable Z_i for unit i has a distribution that depends on covariates $\mathbf{x}_i$ and parameters $\boldsymbol{\theta}$. If there exist functions

$$w_i(Z_i|\mathbf{x}_i, \boldsymbol{\theta}),$$

independently and identically distributed, following a distribution that does not depend on unknown parameters, then

$$\hat{\epsilon}_i = w_i(Z_i|\mathbf{x}_i, \hat{\boldsymbol{\theta}})$$

can fulfill the role of residuals.

Suitable functions for this task can be found using the general result that, if Y is a random variable with distribution function $F(.)$, then the random variable

$$V = F(Y) \sim U(0, 1).$$

It follows that the random variable $U = -\ln(1 - F(Y))$ has pdf

$$g(u) = e^{-u}, \; u > 0,$$

which is the exponential distribution with parameter 1. Since

$$S(t) = 1 - F(t),$$

the residuals in a lifetime data model could be the values

$$-\ln \hat{S}(t_i) = \hat{H}(t_i) = \hat{\epsilon}_i$$

where $\hat{S}(.)$ and $\hat{H}(.)$ are estimates of the survival function and the baseline cumulative hazard function, respectively. In the case of a parametric model such as Weibull regression, $\hat{H}(.)$ follows simply from the estimation of the parameters of the model. These Cox-Snell residuals are regarded as very useful in parametric models.

When an observation is right-censored, then

$$1 - \ln \hat{S}(t_i)$$

is usually used as that observation's residual. The reason for this is as follows. Since the right-censored observation t_i is less than the unknown true value, $-\ln \hat{S}(t_i)$ is likewise less than it should be. The difference between $-\ln \hat{S}(t_i)$ and its true value is similar to a residual lifetime (see section 1.9) and is a random variable that follows the exponential distribution with parameter 1. Consequently, its expected value is 1, and we add on this value in order to estimate the residual that would have been obtained if the observation had not been censored. Making this adjustment for the censored observations, the set of residuals can be examined in a probability plot against the exponential distribution with parameter 1.

EXAMPLE 1.3.– *To illustrate the use of residuals, we fit a Weibull regression model to the set of experimental data given in Table 1.3. These are the failure times of glass capacitors in a* 4×2 *factorial experiment (four levels of voltage, two temperatures) with eight replications. Note that "Type II" censoring was applied: the experiment at each temperature/voltage combination ran until four of the eight units had failed. The remaining four were right censored at that time. The original analysis fitted exponential distributions with a guarantee parameter. (A guarantee parameter is in effect a minimum possible lifetime. The exponential distribution modified in this way has pdf* $f(t) = \lambda \exp(-\lambda(t-\tau))$ *with* $t \geq \tau$*.)*

We fit the Weibull regression model to the data including the right-censored observations, with covariates voltage V and temperature T. We treat V as a quantitative measurement without transformation. Since T takes only two values, it makes no real difference whether we treat it as quantitative or categorical. Fitting is by maximum likelihood, and likelihood ratio tests confirm that both V and T should be included in the model.

Figure 1.5 shows the probability plot, against the exponential (1) distribution, of the Cox-Snell residuals (corrected for right censoring where necessary) from the analysis that includes both covariates V and T. The plot is reasonably close to the expected straight line.

	Applied voltage			
Temperature	200	250	300	350
170^oC	439	572	315	258
	904	690	315	258
	1092	904	439	347
	1105	1090	628	588
	1105*	1090*	628*	588*
	1105*	1090*	628*	588*
	1105*	1090*	628*	588*
	1105*	1090*	628*	588*
180^oC	959	216	241	241
	1065	315	315	241
	1065	455	332	435
	1087	473	380	455
	1087*	473*	380*	455*
	1087*	473*	380*	455*
	1087*	473*	380*	455*
	1087*	473*	380*	455*

Table 1.3. *Zelen's data from life tests of capacitors (lifetimes in hours) [ZEL 59]. Asterisks denote right-censored observations*

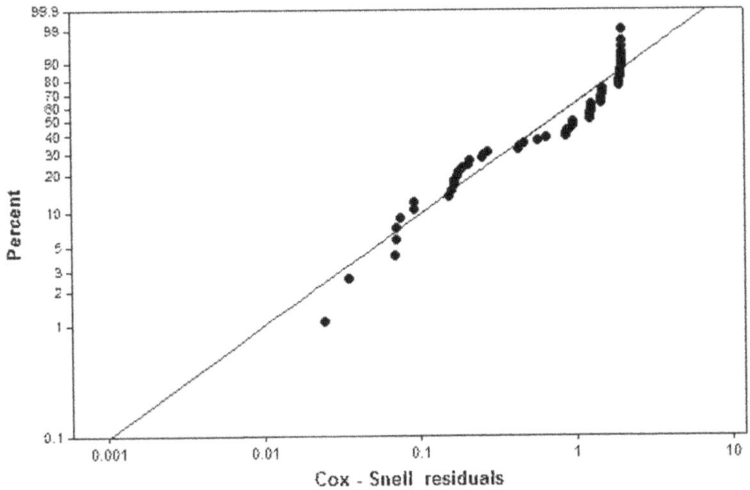

Figure 1.5. *Probability plot of corrected Cox-Snell residuals from Weibull regression model fitted to the data of Table 1.3*

1.17. Cured fraction or long-term survivors

As already noted in the discussion of the Gompertz distribution in section 1.4, it is a characteristic feature of lifetime data - present in most studies - that the data are incomplete, in the sense that some of the units under study have not experienced the event. Therefore, they contribute right-censored lifetimes. Very often, this happens because it is not feasible to allow data collection to continue until all units have failed, because that could take years in contexts such as studies of highly reliable machines or human survival. However, it is implied by the form of the basic models for lifetime data analysis that every unit is susceptible to failure. Therefore, the event would have been recorded for every unit if only the observation could have gone on long enough.

There are many contexts, however, in which it is possible that not every unit is in fact susceptible to failure. The obvious examples come from medical studies of the time from the end of treatment until relapse or death from the disease. If a complete cure is a possibility, then individuals who have been cured by the treatment are free of the disease and therefore will not relapse and will not die from the disease. Having been cured, they are no longer susceptible. The proportion of the population that is no longer susceptible after treatment is called the cured fraction. Other terminologies are immunes and - recognizing that, in practice, no lifetime is infinite - long-term survivors.

If there is a cured fraction, then it follows that the usual condition on the survival function $S(t) \to 0$ $(t \to \infty)$ does not hold: this was also remarked in the discussion of the Gompertz distribution. Sometimes, this can be seen in the appearance of the estimated survival function. Figure 1.6 shows an example from a study of the time to graduation of 15,541 undergraduate students in a Greek technical university (see Caroni [CAR 11b]). Starting from the minimum duration of studies of five years, the number of surviving students (i.e. those who have not graduated yet) falls steeply for about two years, but much more gradually thereafter. Rather than tending towards zero, it looks as if $S(t)$ tends towards a limit of the order of 10%. (In fact, the analysis gives an estimate of 12.4% with a standard error of 0.4.) This is a significant proportion of the student intake, and it should be represented somehow in the model that describes the data.

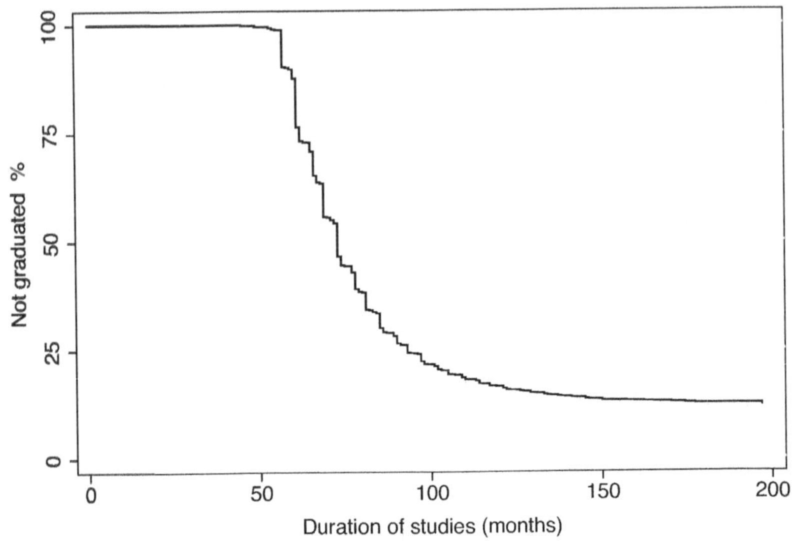

Figure 1.6. *Proportion of students who have not yet graduated, by time since commencement of studies (15541 students in a Greek technical university [CAR 11b])*

The credit for the first attempt to tackle the issue of the presence of a cured fraction is usually given to Boag [BOA 49] who proposed a mixture model with survival function:

$$S(t) = (1 - \pi) + \pi S_0(t),$$

where π is the proportion of the population that is susceptible and $1 - \pi$ the proportion that is non-susceptible. S_0 is the survival function among those who are susceptible, which, in Boag's application to data on breast cancer, is given by the log-normal distribution. Mixture models are also known as split-population models [CHU 91].

Many applications on these lines have appeared (see Maller and Zhou [MAL 96]). A large number of them concern the follow-up of disease, where the concepts of cure and immunity have clear meanings. There are also applications in the social sciences, for example, the study of recidivism (see for example, [CHU 91]): how long until someone released from prison re-offends? In that case, however, the mixture model is open to question. Is it plausible to suppose that everyone can be firmly characterized at the moment of release as either susceptible or immune to re-offending? It seems more realistic to suppose that re-offending, or avoiding re-offending, is an outcome of what happens after the release from prison, rather than being a fate that is already fixed at the moment of release. Similarly, in the analysis of the time until a student's graduation, it is hard to accept a model that characterizes some students from the beginning of their studies as ones who will never graduate. It is much more believable that failure to complete studies arises because of what happens along the way: other life events (marriage, parenthood); fading interest in the subject of study; finding a job that leaves no time for studies, and so on. Farewell [FAR 82, FAR 86] cautioned against the use of mixture models unless there was a clear scientific basis for the existence of an immune proportion.

As already noted in the discussion of the Gompertz distribution, we will see in the following chapter how it is possible for long-term survival to arise as a feature of a model that describes lifetimes, without the need to split the population into groups as in the mixture model.

1.18. Frailty

Covariates are introduced into a statistical model in order to account for heterogeneity between units. However, it is often doubtful that the available covariates are sufficient to represent the heterogeneity completely. There may be other covariates that should ideally have been taken into account, but either were not recorded for some reason, or could not be recorded. An example that is often given of the latter is a genetic factor that is presumed to affect a patient's outcome but cannot be measured.

This unobserved heterogeneity may be introduced into the model by supposing that it can be represented by an individual random effect, specifically each individual's value λ of an unobserved non-negative random variable Λ, which in survival analysis is called the frailty (see Vaupel *et al.* [VAU 79]). The effect of frailty enters the model in a similar way to the effect of covariates. Thus, in a PH context, the hazard function for this individual becomes

$$h(t|\lambda) = \lambda h_0(t),$$

where h_0 is the baseline hazard function as usual. Observed covariates may be included in the same model. Distributions of Λ that are commonly assumed include the gamma distribution (as in the example below), the inverse Gaussian distribution and the positive stable distribution (see Hougaard [HOU 95]).

It is particularly important to allow for all sources of heterogeneity in hazard-based modeling. Without suitable adjustment, a strong selection effect operates: individuals with high frailty λ will tend to die first, leaving a population with relatively low frailty and therefore lower hazard. This gives the impression of a decreasing hazard rate over time; however, this describes the study cohort and should not be misinterpreted as a decline in the risk faced by an individual member of that cohort. The fact that the hazard rate depends on these selection effects as well as on any actual variation in individuals' risk means that, according to Aalen, "the hazard rate is a rather more obscure concept than one should wish, and must be interpreted with great caution" [AAL 94, p. 227]. Selection effects are discussed in detail by Vaupel *et al.* [VAU 79] and Vaupel and Yashin [VAU 85], among others.

However, it should be noted that the frailty term in a PH regression model cannot be used as a handy means of gathering up all the variation that is not accounted for by the measured covariates, including the heterogeneity that would have been accounted for by the unmeasured covariates if they had been available. This is because omitting relevant covariates always results in the attenuation of estimates for the covariates that have been included, as remarked in section 1.15. It has been argued by Hougaard *et al.* [HOU 94] and Keiding *et al.* [KEI 97] that from this point of view, it is preferable to use an AFT model rather than a PH frailty model.

Frailty can also be a very useful device in the analysis of multivariate survival data. Multivariate data can arise when units fall into groups, and common (although unmeasured) factors are expected to be affecting each member of the group: for example, a pair of human twins forms a group of this kind (see Hougaard *et al.* [HOU 92]). This correlation may be represented by each group member having the same value λ of the frailty, even though they possibly differ in values of the measured covariates. Correlation would also be expected between repeated times-to-event measured on the same unit, and the device of introducing an individual random effect may therefore also be useful for recurrent events data (see section 4.5).

EXAMPLE 1.4.– *Suppose that the basic lifetime distribution is Weibull with pdf*

$$S_0(t) = \exp\{-(t/\alpha)^\eta\}$$

and the frailty distribution is a gamma distribution

$$g(\lambda) = \frac{1}{\Gamma(\nu)}\lambda^{\nu-1}e^{-\lambda}$$

with shape parameter ν *and scale parameter 1. Because* $S(t|\lambda) = S_0(t)^\lambda$, *the unconditional survival or reliability function of the lifetime is*

$$\begin{aligned} S(t) &= \int_{\lambda=0}^{\infty} \frac{\lambda^{\nu-1}e^{-\lambda}e^{-\lambda u}}{\Gamma(\nu)}\, d\lambda \quad [where\ u = (t/\alpha)^\eta] \\ &= \int_{\lambda=0}^{\infty} \frac{\lambda^{\nu-1}e^{-\lambda(1+u)}}{\Gamma(\nu)}\, d\lambda \\ &= \frac{1}{(1+u)^\nu}\int_{\lambda=0}^{\infty} \frac{(1+u)^\nu\lambda^{\nu-1}e^{-\lambda(1+u)}}{\Gamma(\nu)}\, d\lambda \\ &= \frac{1}{(1+u)^\nu} \\ &= \frac{1}{\{1+(t/\alpha)^\eta\}^\nu} \end{aligned}$$

since the integral integrates the pdf of a gamma distribution with parameters $1+u$ *and* ν *over its entire range, and therefore equals one. The distribution with this survival function is known as the Burr distribution.*

1.19. Models for discrete lifetime data

The models that have been discussed so far in this Chapter assume that the time measurement T is a continuous random variable. This is usually true, but two other possibilities need to be at least mentioned, although their relative lack of practical importance can be judged by the small number of pages allotted to them even in such a comprehensive text as Lawless [LAW 03]. One of these possibilities is that the "time" variable is inherently discrete; the other is that it is a continuous measurement that has been grouped into categories.

Inherently discrete measurements of the time until an event occurs may correspond to an operational time such as how often a machine has been used. Another example is how many times something has been attempted before success, or for how many semesters a student has been enrolled until obtaining a degree. Even inherently discrete time variables, however, may often be satisfactorily treated as continuous in order to gain access to the rich array of models for continuous times, few of which have any counterpart in discrete time. The main exception is when the lifetime is short. For a simple analogy, a standard model for a count variable is the Poisson distribution, which can often be adequately approximated by a continuous distribution, but certainly not when the parameter value implies that the count variable typically takes small integer values.

Not many useful discrete distributions are available for modeling lifetimes. The simplest one is the familiar geometric distribution that can model the time to the first success in a sequence of independent Bernoulli trials with constant probabilities of success p and failure $q = 1 - p$:

$$P(T = t) = q^{t-1}p, \;\; t = 1, 2, 3, \ldots$$

For a discrete distribution, the hazard rate at time t is defined as the conditional probability of failure at this moment, given survival so far:

$$h_t = P(T = t | T \geq t) = p_t / S_t,$$

where p_t is the probability distribution and $S_t = P(T \geq t)$. For the geometric distribution, $S_t = q^{t-1}$, therefore $h_t = p$, which is constant for all t. Consequently, the geometric distribution can be regarded as the discrete equivalent of the continuous exponential distribution in that it preserves the property of constant hazard. There have been attempts to define equivalents of other well-known lifetime distributions, such as a discrete Weibull distribution. Since the exponential distribution's survival function is

$$S(t) = e^{-\lambda t} = (e^{-\lambda})^t$$

and the Weibull's is

$$S(t) = e^{-(t/a)^\eta} = (e^{-\lambda})^{t^\eta},$$

where $\lambda = \alpha^{-\eta}$, a "discrete Weibull distribution" could be defined by the survival function

$$S_t = q^{(t-1)^\beta}, \; t = 1, 2, \ldots$$

(see Nakagawa and Osaki [NAK 75]).

Further extensions and applications of these ideas are so few that it is not worthwhile pursuing them here. In particular, adapting the regression models presented in the preceding sections from continuous time to discrete time is generally difficult. It is awkward to define a time-transformation method equivalent to AFT that observes the restriction of the time variable to a set of specific discrete values. PH models can be used, but need to be adapted so that the hazards, which are probabilities in the context of discrete time, correctly observe the restriction to the $(0, 1)$ range. (Continuous hazard rates can take any non-negative value.) On the other hand, a PO specification is quite natural for discrete times. In general, the possible models are rather close to generalized linear models (see Lawless [LAW 03]).

Grouped and interval censored data

The second possible form of discrete data arises when all the observations have been recorded in the same intervals. (Grouping may also be used for easier presentation of a large dataset, but if the original observations are available, they should be used in the analysis.) A familiar example of such grouped data is the life table, showing the declining size of a population or cohort year-by-year. Life tables have formed part of the statistical literature for 500 years. Discussion and references on their analysis can be found in Lawless [LAW 03]. An issue that has to be considered is the handling of censored data if the time of censoring is not known exactly, but only known to lie within an interval.

A related form of data arises from interval censoring [SUN 06]. This again means that observations of a continuous time variable have not been recorded exactly, but in intervals. However, the intervals are not necessarily the same throughout the dataset. (It is true of course that it can be claimed that all our apparently continuous measurements are in fact interval censored, with intervals corresponding to the accuracy of our measuring and recording - nearest minute, nearest hour, etc. However, this is rarely a problem; otherwise we would never be able to use the basic tools of statistical analysis.) Grouping of times arises most commonly because individuals are not being observed continuously but only at specific times. Suppose in general that unit i is inspected at a prespecified sequence of times (e.g. a person in a study has monthly appointments at a clinic) until it is found at one of these inspections that the unit has failed (the event of interest has occurred) during the time that has elapsed since the previous inspection. The information available on the lifetime T_i of this unit is $U_i < T_i \leq V_i$, where $(U_i, V_i]$ denotes the interval in which the failure took place. For a unit that had already failed by the time of the first inspection after entering the study, $U_i = 0$ and the observation is left censored. If a unit is still operating at the final inspection before the study is terminated, then $V_i = \infty$ and the observation is right censored. Clearly, the probability of failure in the interval $(U_i, V_i]$ is $F(V_i) - F(U_i)$, where F is the distribution function of lifetimes, with $F(0) = 0$ and $F(\infty) = 1$ as usual. Therefore, the likelihood of the data is simply

$$L = \prod_i \left[F(V_i) - F(U_i)\right]. \quad [1.9]$$

Note that for a fully parametric model - including regression specifications - this likelihood can be maximized with the same effort as is required for maximizing the likelihood [1.3]. In this respect, interval censoring presents no analytical difficulty whatsoever. As mentioned for grouped data above, however, care is required in formulating the problem if any observations were censored within intervals (e.g. after some point, a patient in the study failed to turn up for further appointments) or when the next inspection time is not prespecified or even determined independently of the lifetime process, but instead depends in some way on the unit's condition at the previous inspection. Further details can be found in Lawless [LAW 03].

However, it should be pointed out that, in contrast to fully parametric models, the maximization of [1.9] becomes notably difficult for the semi-parametric Cox PH model, because the remarkable simplification that is provided by the reduction to partial likelihood does not work for interval-censored data. (This is because the unspecified baseline hazard $h_0(t)$ does not cancel out.) Consequently, this basic tool of many medical investigators is unavailable for a research design that they quite commonly use. Some approaches to this problem are reviewed by Caroni [CAR 11a].

New methods have been proposed recently by Sun *et al.* [SUN 15] and Wang *et al.* [WAN 16].

One special case of interval-censored data that is, however, of major importance and consequently has largely developed its own literature is current status data. This arises when there is only one inspection of each unit. This means that all that is known for each unit is the inspection time and whether or not the unit had failed at this time. Every observation must be either left censored (if the unit has already failed) or right censored (not yet failed). For reviews of the topic, see, for example, Diamond and McDonald [DIA 92] and Jewell and van der Laan [JEW 03]. It is interesting to note that common practice in biostatistics involves some contradiction in handling these data. Suppose that all units are inspected after the same time, t_0. Given the binary dependent variable (failed/not failed) and a covariate vector $\mathbf{x}$, the routine methodology calls for fitting a logistic regression model, so that the Bernoulli probability $\pi_{\mathbf{x}}$ of failure is modeled as depending on $\mathbf{x}$ through the logit link function

$$\ln\left\{\frac{\pi_{\mathbf{x}}}{1-\pi_{\mathbf{x}}}\right\} = \boldsymbol{\beta}'\mathbf{x}.$$

However, the same practitioners might routinely use the Cox PH model in the same research area if they had lifetimes recorded exactly. Under PH,

$$\begin{aligned} 1-\pi_{\mathbf{x}} &= S(t_0|\mathbf{x}) \\ &= \exp\left\{-e^{\boldsymbol{\beta}'\mathbf{x}} H_0(t_0)\right\} \end{aligned}$$

for baseline cumulative hazard function $H_0(t)$. However, $H_0(t_0)$, being the same for all units, can be absorbed into the linear predictor $\boldsymbol{\beta}'\mathbf{x}$, giving

$$1-\pi_{\mathbf{x}} = \exp\left(-e^{\boldsymbol{\beta}'\mathbf{x}}\right)$$

which can be rewritten as

$$\ln\left(-\ln(1-\pi_{\mathbf{x}})\right) = \boldsymbol{\beta}'\mathbf{x}.$$

In other words, the generalized linear model for current status data that corresponds to the Cox PH model should not use the logit link function, but the complementary log-log link function instead. Using the logit link (logistic regression) matches the use of the PO model for exactly observed lifetimes, not PH. Of course, as is well known, the logit and complementary log-log transformations are so similar over most of their range that the practical importance of the discrepancy that we have noted is minimal.

1.20. Conclusions

This introductory chapter has aimed to present, at least in outline, the main ways of approaching the regression analysis of lifetime data; that is, how covariates that influence lifetimes can be included in an empirical model. As seen, the most important ways (at least, in terms of the frequency of their application) are by means of the PH model - often equated with Cox's semi-parametric version of the model - and by the AFT model. In addition, the PO model is sometimes seen, and occasionally the additive hazards model, but the others are hardly ever seen. As noted, the great appeal of Cox's model is attributable to the apparent ease of interpretation of the regression coefficients and also to its semi-parametric nature, which avoids the need to specify the parametric form of the baseline hazard function. Although the PH model was initially motivated in section 1.4 by appeal to the device of making a Weibull distribution's parameters depend on the covariates - as is widely done in statistical modeling - the consequence that hazard functions of different units are proportional to each other is a very attractive property. Subsequently, as seen in several sections of this chapter, other "proportional" models have been defined in terms of other functions derived from the distribution of lifetimes, such as the MRL. Because of the need to compete with the predominant semi-parametric Cox model, the promoters of these alternatives have put much effort into their semi-parametric estimation. In fact, in all lifetime data models with non-informative right censoring, fully parametric modeling offers no difficulty in principle. All that is required is to state the model, write down the likelihood for the uncensored and right-censored cases and maximize it numerically.

The AFT model occupies a rather different position. First, it can be written in the form of a familiar regression model, with the logarithm of lifetime as dependent variable and with a suitable error distribution replacing the normal. Second, it is generally used in the fully parametric form, which seems appropriate for its association with experimental data and other reliability data. However, as noted earlier in this chapter, because of its various desirable properties, it should be more widely used in biostatistical applications as various authors have pointed out.

Neither the AFT nor the PH model includes a representation of the process by which the covariates act upon lifetimes. The same applies to the other regression models that have been mentioned in this chapter. The major objective of this book is to present an alternative class of models that sets out to model the lifetime as the observed outcome of some underlying process. Modeling the mechanism may lead to a more satisfactory model offering greater scope for scientific insight than is possible from strictly empirical models.

2

First Hitting Time Regression Models

2.1. Introduction

The concept of a "model" means different things to different people, and its usage in statistics does not always match with what is encountered in many other fields. Its basic statistical meaning is very narrow and technical: "a statistical model is a probability distribution constructed to enable inferences to be drawn or decisions made from the data" (see Davison [DAV 03, p. iv]). All that is demanded of a statistical model in this sense is that its numerical behavior should match the numerical properties of the data with sufficient accuracy in order to support these uses. In particular, a statistical model does not necessarily bear any relation to the physical process that generated these data. An example is the linear regression model, $y = \beta'\mathrm{x} + \epsilon$. There is no requirement that a predictor x should actually contribute toward determining the value of the dependent variable y in the real world, merely that there is a statistical correlation that enables x to be used in predicting y. This is an empirical model in the sense of Cox [COX 90].

However, more widely in the world of science, a model would generally be understood to be a representation of the physical process. As an extreme case, in the biomedical sciences, we encounter "animal models": a disease or other process is studied in a non-human animal in order to shed light on the process in humans. More often, the meaning would be a mathematical model. This represents the physical system symbolically, commonly using devices such as differential equations to represent changes in quantities in one part of the system as a result of the changes in another part. Usually, much simplification is involved, but the intention is that the components of the mathematical model mimic in some way what is happening in the physical system, and that the parameters of the mathematical model will be physically interpretable. Models like these do appear in statistics, and they may be preferable to the purely empirical model. Cox says that "the most appealing models are those that connect directly with subject-matter considerations" [COX 90, p.169], which he labels as "substantive models".

The bulk of lifetime data analysis uses empirical models. One can find many published articles that compare the fits of different lifetime distributions to data without offering any comment on which, if any, of them describes the process. Cox's semi-parametric proportional hazards regression model is a purely descriptive empirical model and may be inferior to other models for that reason, as Cox himself is quoted as saying: "Of course, another issue is the physical or substantive basis for the proportional hazards model. I think that's one of its weaknesses, that accelerated life models are in many ways more appealing because of their quite direct physical interpretation, particularly in an engineering context" [REI 94, p. 450]. A desire for more substantive models than those based on the hazard rate is expressed strongly by Aalen and Gjessing [AAL 01]. They emphasize that the event of interest is the end point of a developing process, which should not be ignored, even if to a large extent it is unknown. Analyzing lifetime data in terms of the process may offer much insight into what is going on. The proposed approach is through the representation of the lifetime as the outcome of an underlying stochastic process, with the event occurring when the process reaches a boundary or threshold for the first time - in other words, the lifetime is seen as the first passage time of the stochastic process to this boundary.

This was not a new idea. An earlier review by Singpurwalla [SIN 95] had also promoted the stochastic process-based approach to failure modeling, with emphasis on engineering applications rather than the biomedical field that Aalen and Gjessing had chiefly in mind. The importance of Aalen and Gjessing's contribution was to awaken interest in alternatives to the proportional hazards regression approach that had so rapidly come to predominate in biomedical applications. Thus, their paper focuses on the understanding of the hazard rate that can be gained by modeling a lifetime as a first passage time to some threshold in an underlying stochastic process. See also Chapter 10 of Aalen *et al.* [AAL 08].

A simple example of this approach through stochastic processes is provided by Markov modeling of the progress of infection with human immunodeficiency virus (HIV) up until death from acquired immunodeficiency syndrome (AIDS). Longini *et al.* [LON 89] suggested a time-homogeneous five-state Markov process, with four transient states (1 = HIV infected but antibody negative; 2 = antibody positive but asymptomatic; 3 = symptomatic; 4 = clinical AIDS) and a final absorbing state 5 = death due to AIDS. The waiting time in each transient state is exponentially distributed with mean equal to the reciprocal of the transition intensity to the next state, so that the distribution of first passage times from state 1 to state 5 - that is, the lifetime - is the sum of four exponential distributions. This is known as a phase-type distribution. Phase-type models generate lifetimes as the time required to reach the absorbing state of interest (failure, death) in a continuous-time, homogeneous, finite-state Markov chain. The lifetime is thus the sum of the waiting times in the various transient states that must be passed through on the way to absorption. These intermediate states are latent or unobservable in many applications. A phase-type model is thus a form of threshold model. A recent review of phase-type models, their history and applications can be found in Slud and Suntornchost [SLU 14]. One further example is Marshall *et al.*'s [MAR 13] application to modeling students' length of stay at university (see section 1.17).

Suppose that the transition intensity matrix of the Markov chain is

$$\mathbf{Q} = \{Q_{ab} : a, b = O, 1, \ldots, K, D\},$$

where the initial state is O, the terminal state is D, and there are K other states. Then, for the absorption time T from O to D,

$$P_{OD}(t) = p(T \leq t) = (\exp(t\mathbf{Q}))_{OD}.$$

Each waiting time within a transient state is exponentially distributed, although generalizations have been proposed (see Huzurbazar [HUZ 99]). A model with many states and transitions has a large number of parameters, but it may be possible to reduce the number by equating some of these. Slud and Suntornchost warn that only models with relatively simple path structures can realistically be fitted stably [SLU 14]. The application that they examine in detail is for breast cancer, in which the model postulates two alternative pathways to the end state (a situation that has been demonstrated by medical research) as well as cure. A six-parameter model, with four transient states on one pathway and one on the other, provides a reasonable fit to the quite complex observed hazard function, and more complicated models were then developed from it.

This application does not explicitly include covariates, although the two pathways do in fact correspond to two subpopulations defined by the positive/negative status of one covariate. There is no difficulty in principle in introducing covariates directly by constructing regression models for transition rates. This has been done, for example, by Faddy and McClean [FAD 99]. However, this introduces more parameters and potentially leads to poor estimation [SLU 14].

In many applications, however, discrete states of the stochastic process cannot be distinguished. The principal example of the approach in that case is the model based on the first passage time of an underlying Wiener process until it reaches a boundary. The Wiener process, defined in the next section, is a continuous-time stochastic process with continuous state space. As will be seen, its first passage times follow the inverse Gaussian distribution. Regression methods in this framework have been particularly developed by Lee and Whitmore ([LEE 06] and many subsequent papers), following earlier contributions by Whitmore.

2.2. First hitting time models

There are two basic components in a first hitting time (FHT) model. The first is the parent stochastic process $\{Y(t),\ t \in \mathcal{T},\ y \in \mathcal{Y}\}$ with $Y(0) = y_0$. As will be seen in the following examples, many types of process are possible, including processes with discrete or continuous states, and monotonic or non-monotonic sample paths. Although in some cases the process $Y(t)$ will be observable, it will more often be an unobservable, latent process.

The second component is a boundary set or threshold, $\mathcal{B} \subset \mathcal{Y}$. In the basic model, this is assumed to be fixed; however, more generally, it could also depend on time $\mathcal{B}(t)$. Assuming that the process starts at time zero outside the boundary set, the FHT (or first passage time) is defined as the time elapsed from zero until the process first enters $\mathcal{B}$:

$$S = \inf \{t : Y(t) \in \mathcal{B}\} .$$

It should be noted that the event of reaching the boundary is not guaranteed to happen in general: for some processes, it is possible that $P(S < \infty) < 1$. As will be seen later, this may be an important feature of the model. Several models within this general framework are presented briefly by Lee and Whitmore [LEE 06], as follows.

Wiener process. The process $Y(t)$ is a random walk in continuous time and space, more specifically a Brownian motion with drift. This example, which will be considered in detail in the following sections, is in fact the major topic of this book. The distribution of FHTs follows the inverse Gaussian distribution. Some early examples of the application of this distribution to modeling lifetimes include Lancaster's study of the duration of industrial strikes [LAN 72], and Eaton and Whitmore's analysis of length of stay in hospital [EAT 77]; see also Whitmore [WHI 75]. In a different area, Dennis *et al.* base models of the extinction of endangered species on the Wiener process [DEN 91].

Gamma process. In contrast to a process modeling human health, which ought to allow improvement as well as deterioration, a process that models the degradation of a material (e.g. the growth of a crack) might be required to be monotonic on the grounds that improvements cannot occur (the damage cannot usually mend itself). One suitable model is a gamma process. Let $Y(t) = y_0 - Z(t)$, where $Z(t),\ t \geq 0$ is a gamma process with scale parameter β, shape parameter α and $Z(0) = 0$. The FHT of the zero level in the parent process ($Y = 0$) follows an inverse gamma distribution, defined by the identity $P(S > t) = P\left(Z(t) < y_0\right)$. Gamma processes will be mentioned further in section 2.9.

Bernoulli process. A sequence of Bernoulli trials with probability of success p on each trial can model the discrete time (number of trials) required in order to reach a specified number of successes m. The number of trials carried out would follow the negative binomial distribution. In the general notation, $Y(0) = y_0 = m$ and let $Y(t) = y_0 - B_t,\ t = 1, 2, \ldots$, where $\{B_t\}$ is the preceding Bernoulli process and the boundary set is at zero. This process could model phenomena such as the time required to dispose of a stock of m items, or obtain m completed interviews from a population in a sample survey. It is also used empirically to represent lifetimes without any underlying model: for example, as an overdispersed alternative to a simple Poisson distribution to describe the number of days of hospitalization (see

Carter and Potts [CAR 14b]). This is because the negative binomial distribution also arises as a compound distribution, with a gamma distribution for the parameter of a Poisson distribution - in other words, the idea of frailty (as in section 1.18).

Poisson process. Like the Bernoulli process, this is a model in which the observed event is triggered by the occurrence of the mth event in an underlying stochastic process. Let $Y(t) = y_0 - N_t$, where $N_t,\ t \geq 0$ is a Poisson process with rate parameter λ. The event occurs when $Y(t)$ reaches zero for the first time. The distribution of the time to the event is the Erlang distribution with parameters m and λ:

$$f(t|m, \lambda) = \frac{\lambda^m t^{m-1} e^{-\lambda t}}{(m-1)!},\ t,\ \lambda > 0,$$

which is simply the special case of a gamma distribution with integer shape parameter.

This is a potential model for the time to overall failure of a system of m parallel components with independent and identically distributed exponential lifetimes, each component entering service when the previous one fails.

Markov chain models and *semi-Markov processes*. This example has already been outlined in section 2.1. To the Markov chain set-up, the semi-Markov process adds the random time spent in each state. Lee and Whitmore refer to models and possible applications in the study of cancer risks [LEE 06].

Ornstein-Uhlenbeck process. This is a mean-reverting process, which can be described as a Wiener process with the added feature of a tendency to drift toward a central location, which is its long-term mean. Aalen and Gjessing [AAL 04] first looked at this process in detail as a model for survival analysis, although without describing specific applications. Results for the first passage time distribution were given by Ricciardo and Sato [RIC 88]. See work by Erich and Pennell described in section 2.7.

2.3. First hitting time regression models based on an underlying Wiener process

Suppose that the final event (disease diagnosis, death, breakage, etc.) is the observable outcome of an underlying process. For example, the progressive degradation of a material eventually leads to its breakage; a human being's deteriorating "health status" leads to disease or death when it falls too low. This underlying process may well be only conceptual; in practice, it is unobservable. That is, it is a latent process.

Model the underlying process as a stochastic process $Y(t)$, where t is the time variable (not necessarily clock time) and $Y(0) = y_0 > 0$. Suppose that failure (breakage, death, etc.) takes place at time s, which is the time when the event $Y(t) \leq 0$ occurs for the first time. Thus, the lifetime is the time taken for the process to reach the threshold zero for the first time. Modeling the process as falling from a variable starting point to a fixed threshold (usually at zero) is a natural choice in many contexts, such as representing individuals' health status or the distance between opposing parties in a strike [LAN 72]. However, in other situations, it may be more appropriate to suppose a fixed starting point (probably at zero) and a variable threshold above zero. These are just alternative parameterizations, although with different interpretations. The second parameterization may be the more natural one if failure results from accumulated damage, such as the wear on a machine or the build-up of atherosclerosis leading to coronary heart disease or myocardial infarction [COM 13].

One example of a stochastic process that can be used here is the Wiener process defined as follows (see Cox and Miller [COX 65]):

a) $Y(t)$ has independent increments, so that $Y(t_2) - Y(t_1)$ and $Y(t_4) - Y(t_3)$ are independent for any pair of non-overlapping intervals (t_1, t_2) and (t_3, t_4);

b) for any interval (t_1, t_2),

$$Y(t_2) - Y(t_1) \sim N(\mu(t_2 - t_1), \sigma^2(t_2 - t_1)).$$

Supposing that $\mu \leq 0$, then the lifetime T follows the inverse Gaussian distribution:

$$f(t|y_0, \mu, \sigma^2) = \frac{y_0}{(2\pi\sigma^2 t^3)^{1/2}} \exp\left[-\frac{(y_0 + \mu\, t)^2}{2\sigma^2\, t}\right].$$

(If $\mu > 0$, the event $Y(t) \leq 0$ is not certain to occur. This case will be pursued later.) In our context, the model is over-parameterized (because Y has an arbitrary scale), and we usually take $\sigma = 1$. Although the parameters μ and y_0 have very simple interpretations in terms of the underlying process, they do not have one in terms of the lifetime distribution. The mean lifetime is $y_0/|\mu|$ and the variance is $y_0/|\mu|^3$.

Now introduce the effects of covariates by allowing the parameters μ and y_0 of $Y(t)$ to depend on covariates $\mathbf{x}$. This is saying, for example, that humans differ in their original health status and also in the rate of change of health status. Suitable models are

$$\begin{aligned} \mu &= \boldsymbol{\beta}'\mathbf{u} \\ \ln y_0 &= \boldsymbol{\gamma}'\mathbf{v}, \end{aligned} \qquad [2.1]$$

where the parameters $\boldsymbol{\beta}$ and $\boldsymbol{\gamma}$ are vectors of regression coefficients. The vectors of covariates $\mathbf{u}$ and $\mathbf{v}$ may share common elements, or even be identical. It may sometimes be convenient to write $\boldsymbol{\beta}'\mathbf{x}$ and $\boldsymbol{\gamma}'\mathbf{x}$ for the two linear predictors, with the understanding that some elements of $\boldsymbol{\beta}$ and $\boldsymbol{\gamma}$ may be set equal to zero *a priori*.

A comment on terminology should be made at this point. We will always use the terminology "first hitting time regression" or FHT regression for this model and others that are also based on the first passage of a stochastic process to a boundary state or threshold. Lee and Whitmore ([LEE 06] and elsewhere) generally use "threshold regression". We avoid this term because, although it is clear enough in the literature on lifetime data analysis, it is ambiguous elsewhere. In the econometrics literature in particular, the name of threshold regression has been taken by another method, in which an independent variable has different effects on the dependent variable according to whether the value of the independent variable is below or above a threshold level.

Examples of the application of the above method will be presented later, and issues arising will be discussed then. At this point, it suffices to give the clear illustration of the possibilities offered by the inverse Gaussian FHT regression model that is found in the simple example presented in Lee and Whitmore [LEE 10a]. These data on two groups of kidney dialysis patients were used extensively by Klein and Moeschberger [KLE 03], whose tests indicated that the PH property did not hold. The Kaplan-Meier estimates of survival curves in each group cross over ([KLE 03], their Figure 7.1), which is itself an indication against PH (see section 1.11). Furthermore, it looks as if survival in the percutaneous group does not approach zero, but instead is leveling off at about 80%; a model that includes long-term survivors may be required. Lee and Whitmore fit the inverse Gaussian FHT regression model with a single covariate (the indicator variable indicating group membership: 0 for surgical, 1 for percutaneous) included in linear predictors for both the drift μ and the initial level y_0, with identity and logarithmic link functions as usual. The parameter estimates for the linear predictor of the drift parameter are -0.0959 for the constant term and 0.6377 for the coefficient of the group indicator. Consequently, estimated drift is -0.0959 for patients in the surgical group and 0.542 in the percutaneous group. As the latter is positive, reaching the boundary at zero is not certain. In fact, as discussed in the following section, the probability of not reaching the boundary - and hence appearing to be a long-term survivor - is

$$1 - \exp(-2 \times 1.402 \times 0.542) = 0.78,$$

(where 1.402 is the estimate of y_0 for units in this group), which agrees well with the Kaplan-Meier estimate. On the other hand, the estimated drift in the surgical group is negative; therefore, patients reach the boundary with probability one and long-term survivors do not appear. The fitted survival curves in the two groups exhibit the same characteristics as the two Kaplan-Meier estimates including the crossing ([KLE 03], their Figure 7.1). A particularly persuasive feature of this example is that the presence of apparent long-term survivors in one group but not the other emerges naturally as a feature of the basic model. This is far more satisfactory than trying to represent it by, say, including a cured fraction by means of a mixture model (see section 1.17) in one group but not the other.

2.4. Long-term survivors

As indicated above – and discussed earlier in section 1.17 – it is sometimes desirable for the lifetime model to represent the presence of a proportion of units that will apparently never fail. Depending on the context, these units may be labeled the cured fraction, immunes or long-term survivors. In section 1.4, it was shown how the Gompertz distribution with certain parameter values can represent this feature of the model without having to introduce special components such as in the mixture model of section 1.17. A lifetime model based on an underlying Wiener process shares the same property.

In the parameterization of the previous section, when $\mu > 0$, the drift tends to take the Wiener process away from the boundary. This results in a non-zero probability of never reaching the boundary, that is, of the event never happening. Consequently, when $\mu > 0$ for some or all of the units, a proportion of long-term survivors will appear.

Specifically, the probability that the path of the Wiener process never reaches the boundary when $\mu > 0$ is

$$1 - e^{-2\mu y_0}$$

[COX 65]. Conditionally on reaching the boundary, the inverse Gaussian distribution given above still applies and the formulae for mean and variance hold [WHI 78].

Whitmore discussed this defective inverse Gaussian (DIG) distribution in detail [WHI 79] and applied it successfully to modeling the length of service of employees. The underlying process that he envisaged describes an employee's level of "job attachment". The employee decides to leave when his or her attachment falls to a "separation threshold". This conceptual model is far more satisfactory than a mixture model in this situation, because the timing of eventual departure or remaining in the job are outcomes that are both determined by the path of the underlying process. The same process applies to every individual. The mixture model, on the other hand, conceptualizes each individual as being classifiable at the moment of entry into service as someone who might leave at some time point or someone who will never leave under any circumstances. This is implausible.

The DIG model arising from the Wiener process is discussed further by Balka *et al.* [BAL 09], with various extensions. These include a mixture of the inverse Gaussian with a cured fraction, intended to offer a more flexible alternative to the DIG, although it has the drawback of losing the interpretation of the survival times and the cured fraction as a result of an underlying Wiener process. On the other hand, the fact that the cure rate is a function of the parameters of the survival time distribution of the susceptibles is possibly restrictive in some circumstances. See section 3.3 for the fitting of these models. Other extensions include the mixtures of DIGs and models with random heterogeneity in the parameters of the Wiener process. Furthermore, one interesting model represents the cured fraction as developing by absorption of the Wiener process at a second barrier. A different model with two absorbing barriers - but corresponding to two observable outcomes - is discussed in section 4.6. Applications of some of the models of Balka *et al.* are given in section 6.12.

2.5. FHT versus PH

Although the general idea of using an FHT model of the process underlying the time to an event can lead to various lifetime distributions, depending on the nature of this underlying process, the one found most commonly - at least in areas outside engineering - is the inverse Gaussian distribution arising from an underlying Wiener process; we will focus our attention on this distribution. In this section, we compare its properties to those of distributions that satisfy proportional hazards. The inverse Gaussian is not a PH distribution and neither is it an AFT nor a PO distribution. Extensive general information on the inverse Gaussian distribution can be found in Tweedie [TWE 57], Folks and Chhikara [FOL 78], Chhikara and Folks [CHH 89], Seshadri [SES 97] and elsewhere.

Although the origins of the distribution go back further, the introduction of the inverse Gaussian distribution into the statistical literature is attributed to Tweedie's 1945 paper [TWE 45]. He also supplied the name, which arises from the fact that the cumulant generating function of the distribution is the inverse of that of the Gaussian distribution. (Some writers think the name inappropriate, because they misinterpret "inverse" as "reciprocal".) According to Fries and Bhattacharayya [FRI 83], interest in the inverse Gaussian as a lifetime distribution is justified by (a) its flexibility in that it takes a variety of shapes, (b) the convenient properties of its sampling distributions, which are often analogous to the normal distribution [CHH 77], and (c) the fact that "its derivation from a plausible stochastic formulation of the failure process often provides a physical support to its empirical fit" [FRI 83, p.820]. The first of these points requires some clarification.

The pdf of the inverse Gaussian distribution is usually written as

$$f(t|\theta,\lambda) = \frac{\sqrt{\lambda}}{\sqrt{2\pi t^3}} \exp\left\{-\frac{\lambda}{2\theta^2}\frac{(t-\theta)^2}{t}\right\}, \quad t > 0, \quad \theta, \lambda > 0. \qquad [2.2]$$

In this parameterization, the mean of the distribution is θ and the variance is θ^3/λ. (In fact, the symbol μ is usually used where we have written θ. However, we wish to reserve μ to represent one of the parameters of the alternative FHT formulation of the inverse Gaussian distribution.)

The survival function $S(t) = P(T > t)$ is

$$S(t) = \Phi\left[\sqrt{\lambda/t}(1 - t/\theta)\right] - e^{2\lambda/\theta}\,\Phi\left[-\sqrt{\lambda/t}\,(1 + t/\theta)\right],$$

where $\Phi(.)$ is the standard normal distribution. The shape of the hazard function $h(t) = f(t)/S(t)$ of the inverse Gaussian distribution was analyzed in detail by Chhikara and Folks [CHH 77], who showed that it always has the increasing-then-decreasing-failure-rate (IDFR) shape, for any θ and λ. Various examples are shown in Figure 2.1.

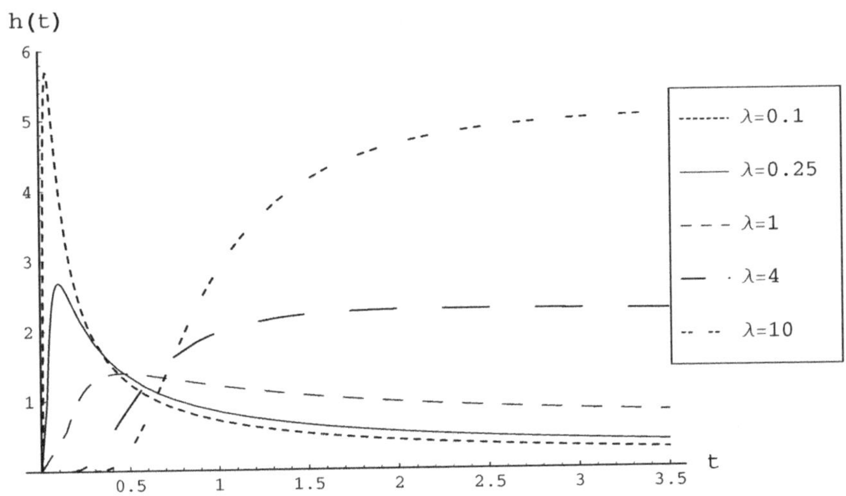

Figure 2.1. *Hazard function $h(t)$ of the inverse Gaussian distribution for various values of λ with θ= 1, from Stogiannis et al. [STO 11]*

The modal value of the hazard function occurs at t^* which is found by solving

$$h(t) = \frac{3}{2t} + \frac{\lambda}{2\theta^2} - \frac{\lambda}{2t^2}.$$

	Modal time	Survival	Hazard		
λ	t^*	$S(t^*)$	$h(t^*)$	$h(\infty)$	$h(t^*)/h(\infty)$
0.1	0.039	0.879	5.70	0.05	114
0.25	0.101	0.854	2.68	0.125	21.4
0.5	0.210	0.804	1.73	0.25	6.93
0.75	0.329	0.744	1.47	0.375	3.92
1	0.457	0.674	1.39	0.5	2.78
1.5	0.734	0.516	1.40	0.75	1.87
2	1.035	0.353	1.52	1	1.52
3	1.684	0.112	1.86	1.5	1.24
4	2.363	0.020	2.28	2	1.14
5	3.052	0.002	2.72	2.5	1.09
7.5	4.777	3.7×10^{-7}	3.90	3.75	1.04
10	6.487	1.3×10^{-12}	5.11	5	1.02

Table 2.1. *Time t^* at which the mode of the inverse Gaussian hazard function $h(t)$ occurs, values of hazard function h and survival function S at that time, and asymptotic value $h(\infty)$ of the hazard. IG parameter $\theta = 1$ (expanded from Stogiannis et al. [STO 11])*

Table 2.1 gives t^* for various values of λ (with θ=1) and also shows $S(t^*)$. Because $\theta = 1$, the expected lifetime is unity. It can be seen that for small values of λ/θ, the peak of the hazard function occurs early and a high proportion (nearly 0.9) of individuals survives beyond t^*. Consequently, some early failures occur but after this relatively short time, the hazard function declines. On the other hand, for large values of λ/θ, nearly all the individuals have experienced the event by the time the hazard function reaches its peak, so that effectively the situation is one of increasing failure rate (IFR). Therefore, the inverse Gaussian hazard function, even though it technically always has a maximum, can in practice behave rather like an IFR or a decreasing failure rate (DFR). For this reason, it can fairly be described as being flexible.

The hazard function of the inverse Gaussian distribution approaches the limit $\lambda/2\theta^2$ as $t \to \infty$ [CHH 77]. We now examine the corresponding result in terms of the usual parameterization of the inverse Gaussian distribution when it is derived as an FHT model. This parameterization [LEE 06] is

$$f(t|\mu, \sigma^2, y_0) = \frac{y_0}{\sqrt{2\pi\sigma^2 t^3}} \exp\left\{-\frac{(y_0 + \mu t)^2}{2\sigma^2 t}\right\}, \quad t > 0,$$

$$-\infty < \mu < \infty, \ \sigma^2 > 0, \ y_0 > 0,$$

where y_0 is the starting level of the underlying Wiener process and μ is the drift parameter of the process, both of which in general depend on covariates. (Again, we note that we usually set $\sigma = 1$.) These parameters correspond to those used above in [2.2] by

$$\lambda = y_0^2, \ \ \theta = y_0/|\mu|,$$

therefore the limit of $h(t)$ is $\mu^2/2$. Consequently:

a) the limit of $h(t)$ is independent of the starting level of the process;

b) for large t, the ratio $h(t|\mathbf{x}_1)/h(t|\mathbf{x}_2)$ for two units with covariate vectors $\mathbf{x}_1$ and $\mathbf{x}_2$ is independent of t.

The first observation indicates that initial differences between the units lose their importance over time, in the sense that two units with the same drift μ but different starting levels y_0 tend toward equal hazards as time progresses. This is in contrast to PH under which any initial differences continue to have the same effect forever. This is an important advantage of this FHT model over PH, which might make it more realistic in many applications. However, the second observation implies that, for large enough t, PH holds (with constant of proportionality depending only on the drift of the process, and not on the starting conditions). Consequently, hazards for a covariate that affects only the starting level of the process converge over time, but hazards for a covariate that affects the drift are eventually proportional. As noted in section 1.8, hazards in proportional odds models also converge over time [BEN 83a].

Lee and Whitmore have studied the connection between FHT and PH regression models in detail [LEE 10a]. They showed that PH models can be obtained from FHT models by varying the time scale or by varying the boundary. In this sense, they regard FHT models as more general and PH as a special case. See section 5.3.

2.6. Randomized drift in the Wiener process

Extensions to the FHT regression model based on a Wiener process will be described chiefly in Chapter 4. However, it is convenient to mention here one straightforward extension that involves no additional complexity. This is the introduction of a random component into the value of an individual's drift parameter, as described in detail by Aalen *et al.* ([AAL 08], their Chapter 10) and mentioned earlier by Aalen [AAL 94, AAL 01] and in a slightly different form by Whitmore [WHI 86b].

The probability density of the Wiener process at time t is

$$\psi_t(y) = \frac{1}{\sigma\sqrt{2\pi t}}\left\{\exp\left[-\frac{(y-c+\mu t)^2}{2\sigma^2 t}\right]\right.$$
$$\left. - \exp\left(\frac{2c\mu}{\sigma^2}\right)\exp\left[-\frac{(y+c+\mu t)^2}{2\sigma^2 t}\right]\right\} \qquad [2.3]$$

[COX 65]. A random element is introduced into the drift by supposing that an individual's value of this parameter is normally distributed with variance τ^2 about the mean $-\mu$ which may depend on covariates in the usual way. Integrating the density [2.3] over the distribution of the drift parameter gives the following density for the transition from c to y in time t

$$\psi_{R,t}(y) = \frac{1}{\sqrt{2\pi(t^2\tau^2+t)}}\left\{\exp\left[-\frac{(y-c+\mu t)^2}{2(t^2\tau^2+t)}\right]\right.$$
$$\left. - \exp\left(2c\mu + 2c^2\tau^2\right)\exp\left[-\frac{(y+c+2ct\tau^2+\mu t)^2}{2(t^2\tau^2+t)}\right]\right\},$$

where the subscript R indicates that the drift parameter has been randomized.

The probability of not having been absorbed by time t is

$$S_R(t) = \Phi\left(\frac{c-\mu t}{\sqrt{t^2\tau^2+t}}\right)$$
$$- \exp\left(2c\mu + 2c^2\tau^2\right)\Phi\left(\frac{-c-2ct\tau^2-\mu t}{\sqrt{t^2\tau^2+t}}\right) \qquad [2.4]$$

[AAL 94]. Differentiating this gives the density of the time to absorption

$$f_R(t) = \frac{c}{\sqrt{2\pi}} \frac{1}{t\sqrt{t^2\tau^2 + t}} \exp\left[-\frac{(c-\mu t)^2}{2\left(t^2\tau^2 + t\right)}\right], \qquad [2.5]$$

which alternatively can be obtained by integrating the inverse Gaussian distribution of lifetimes over the distribution of the drift parameter [WHI 86b, AAL 94]. The possibility that the process will never reach the boundary carries over from the unrandomized Wiener process. In fact, the probability of never being absorbed is

$$\lim_{t\to\infty} S_R(t) = \Phi\left(-\frac{\mu}{\tau}\right) - \exp\left(2c\mu + 2c^2\tau^2\right)\Phi\left(-2c\tau - \frac{\mu}{\tau}\right).$$

Substituting the survival function [2.4] and density [2.5] into the likelihood [1.3], with appropriate expressions for c and μ in terms of covariates, this model can be fitted with no more complexity than is involved in fitting the unrandomized process. A computer program and an illustrative application are mentioned in section 3.5.

It is also possible to randomize the starting level of the process. If a gamma distribution

$$g(y) = \mu^{k+1} y^k e^{-\mu y} / \Gamma(k+1)$$

is taken for the starting level y, then integrating the inverse Gaussian lifetime distribution with $\sigma = 1$ over this gives the unconditional lifetime distribution

$$f(t) = \left(\mu^2/2\right)^{(k+1)/2} t^{(k-1)/2} e^{-\mu^2 t/2} / \Gamma\left((k+1)/2\right),$$

which is another gamma distribution [AAL 08]. Whitmore [WHI 86b], using a parameterization of the inverse Gaussian distribution that describes absorption at a fixed boundary from a fixed starting point at zero, adopted a gamma distribution for the inverse of the variance parameter of the process and a normal distribution for the drift parameter conditional on the value of the variance. Desmond and Chapman developed the part of the model for randomized drift to model the time required to complete a task [DES 93].

2.7. First hitting time regression models based on an underlying Ornstein-Uhlenbeck process

As indicated in section 2.2, the Ornstein-Uhlenbeck (OU) process bears a lot of similarity to the Wiener process and is another potential model for the underlying stochastic process in an FHT lifetime data model (see Aalen and Gjessing [AAL 01, AAL 04] and Lee and Whitmore [LEE 06]). An OU process $Y(t)$ is defined by the stochastic differential equation

$$dY(t) = (a - bY(t))dt + \sigma dW(t),$$

where a, b and $\sigma > 0$ are parameters and $W(t)$ is a standard Brownian motion. The size of the increments in an OU process therefore depends on the current value and, specifically, on the distance between the current value and a/b. The tendency is to drift towards this point, that is,

$$E(Y(t)) = a/b \quad (t \to \infty).$$

The OU process is important as a model of biological processes that exhibit homeostasis, the tendency to revert to an equilibrium point. Many of the functions of an organism have this fundamental feature.

The first attempt to frame the OU as an FHT model and apply it to the analysis of lifetime data was made by Erich and Pennell [ERI 15]. The FHT of the general process from an initial value y_0 to the limit a/b cannot be expressed in closed form. However, an expression does exist in the special case when $a = 0$, $b = 1$ and $\sigma^2 = 2$ (Ricciardi and Sato [RIC 88]). The probability density function of the FHT on the boundary (which is zero with this choice of parameter values) is

$$f\left(t|y_{0i}\right) = y_{0i}\sqrt{\frac{2}{\pi}}\frac{e^{2t}}{\left(e^{2t}-1\right)^{3/2}}\exp\left(-\frac{y_{0i}^2}{2\left(e^{2t}-1\right)}\right),$$

and the survival function is

$$S\left(t|y_{0i}\right) = 2\Phi\left(\frac{y_{0i}}{\sqrt{e^{2t}-1}}\right) - 1.$$

Erich and Pennell call this the Ricciardi-Sato distribution. Its usefulness is apparently limited by the fact that it has only one free parameter (y_0). In particular, it lacks a scale parameter. They therefore propose the introduction of an unknown scale parameter $\alpha > 0$ so that the actual time T recorded in the units of analysis is modeled as $T = T^*/\alpha$, where T^* follows the Ricciardi-Sato distribution. The survival function given above is modified slightly to

$$S\left(t|y_{0i},\alpha\right) = 2\Phi\left(\frac{y_{0i}}{\sqrt{e^{2\alpha t}-1}}\right) - 1,$$

with a corresponding simple change to the density function. This device of transformation from actual time to an operational time is, as will be seen, quite common in the FHT regression modeling literature. To complete the model, a regression structure relating the logarithm of y_0 to covariates is adopted as in the Wiener-based FHT regression model.

One point of interest is the following interpretation of regression coefficients in terms of median survival times. From the expression for S, the median survival time $M_{\mathbf{x}}$ for an individual with covariates $\mathbf{x}$ is

$$M_{\mathbf{x}} = \frac{1}{2\alpha}\ln\left(\frac{y_0^2 + \left[\Phi^{-1}(0.75)\right]^2}{\left[\Phi^{-1}(0.75)\right]^2}\right).$$

If y_0 is sufficiently large, say $y_0 \geq 2$, this is

$$\begin{aligned} M_{\mathbf{x}} &\approx \frac{1}{2\alpha}\ln\left(\frac{y_0^2}{\left[\Phi^{-1}(0.75)\right]^2}\right) \\ &= \frac{1}{\alpha}\left[\gamma_0 + \gamma_1 x_1 + \cdots + \gamma_r x_r - \ln\left\{\Phi^{-1}(0.75)\right\}\right]. \end{aligned}$$

Consequently, a one-unit increase in x_j holding other covariates constant is approximately equivalent to increasing the median survival time by γ_j/α.

The illustrative application of their model given by Erich and Pennell [ERI 15] is to the set of data on malignant melanoma that appears frequently in the literature (see also sections 4.4 and 4.10). Because of the possibility of cure by the initial treatment, a mixture model is adopted with survival function

$$S^*\left(t_i|p_i,y_{0i},\alpha\right) = p_i + (1-p_i)S_{RS}\left(t_i|y_{0i},\alpha\right),$$

where S_{RS} is the survival function of the scaled Ricciardi-Sato distribution given above, and p_i is the probability that subject i has been cured and is therefore not susceptible to the event. The subscript on the cure probability is necessary because it is also modeled as depending on the covariates, through the logit link, as in Xiao *et al.* [XIA 12],

$$p_i = \frac{\exp\{\beta_0 + \beta_1 w_{i1} + \cdots + \beta_q w_{iq}\}}{1 + \exp\{\beta_0 + \beta_1 w_{i1} + \cdots + \beta_q w_{iq}\}},$$

where the covariate vector $\mathbf{w}$ is not necessarily the same as $\mathbf{x}$.

A simulation study showed satisfactory results for the estimation of a correctly specified model. For variously misspecified models, some aspects of the estimation were less satisfactory. In the application to data, the results of fitting the new OU model were compared to those of fitting an FHT model based on the Wiener process. Based on the closeness to Kaplan-Meier estimated survival curves and on the Integrated Brier Score (IBS), which is an index of the inaccuracy of predictions of lifetimes (see Graf *et al.* [GRA 99]), there was no great difference between the two models as far as fit to overall survival was concerned. In the analysis for relapse-free survival, neither model fitted better throughout the range of lifetimes, although the IBS was appreciably less for the OU model. The Wiener process model to which it was compared did not specifically include a cured fraction because of the possibility of a cured fraction emerging naturally from it.

This OU FHT model is interesting as an attempt to get away from the independent increments assumed in the stochastic process that underlies the Wiener-based model. Reservations might be expressed about the fact that its increments, which depend on the current distance from the equilibrium point, are larger for large distances and become smaller as the equilibrium is approached. This seems correct for a model of homeostasis in an organism, where the equilibrium is the desirable state that should be restored quickly. It is less convincing when the limit point is undesirable, as in the zero state of health and consequently death in the above model. It could be equally or more plausible to suppose that a subject whose health is still at a good level tends to remain there (small increments), but once a decline has started, it is likely to deteriorate more quickly (large increments). This is the opposite of what happens in the OU process.

2.8. The Birnbaum-Saunders distribution

The Birnbaum-Saunders distribution [BIR 69] deserves to be mentioned here, as it is another lifetime distribution that arises as an FHT in a model of the underlying process. It represents the failure of a material as the result of the growth of a crack in response to the stress placed upon the material. A simple derivation is as follows. Suppose the material is subjected to repeated cycles of stress and the jth cycle increases the crack size by an amount X_j, distributed with mean μ and variance σ^2 independently of other cycles. The total size after n cycles can therefore be represented as $N(n\mu, n\sigma^2)$. It follows that the probability that the crack size at this point is still less than a critical length ω that would cause failure is

$$P(X \leq \omega) = \Phi\left(\frac{\omega - n\mu}{\sigma\sqrt{n}}\right)$$

and hence, if T is the number of cycles to failure, and ignoring its discrete nature,

$$P(T \leq t) = 1 - \Phi\left(\frac{\omega - t\mu}{\sigma\sqrt{t}}\right) = \Phi\left(\frac{\mu\sqrt{t}}{\sigma} - \frac{\omega}{\sigma\sqrt{t}}\right).$$

The usual parameterization of the distribution is

$$F(t|\alpha, \beta) = \Phi\left\{\frac{1}{\alpha}\left(\sqrt{\frac{t}{\beta}} - \sqrt{\frac{\beta}{t}}\right)\right\}, \; t > 0, \; \alpha > 0, \; \beta > 0,$$

where α is a shape parameter and β is a location parameter (in fact, it is equal to the median). As $\beta = \omega/\mu$, it can be interpreted as the number of cycles to failure that would be required in the absence of random variation. A derivation of the distribution under less restrictive assumptions than those stated above is given by Desmond [DES 85].

Given that both the inverse Gaussian distribution (based on a Wiener process) and the Birnbaum-Saunders distribution can be derived as first passage times in a process with independent normal increments, it is not surprising that there are connections between the two. In fact, Desmond shows that the Birnbaum-Saunders distribution is the equally weighted mixture of an inverse Gaussian distribution and the reciprocal of another inverse Gaussian distribution [DES 86]. Actually, this result was given earlier in the literature by Bartholomew [BAR 73, p. 203], in his discussion of Lancaster's well-known application of the inverse Gaussian distribution in the modelling of the duration of industrial strikes [LAN 72].

A regression model for the Birnbaum-Saunders distribution was developed by Rieck and Nedelman [RIE 91]. This takes the accelerated failure form

$$\ln T = \beta' \mathbf{x} + \epsilon,$$

where the error term ϵ follows the sinh-normal distribution with shape parameter α, location zero and scale equal to 2, denoted by SN(α,0,2). The distribution of the general form SN(α, γ, σ) is

$$F(y) = \Phi\left[\frac{2}{\alpha}\sinh\left(\frac{y-\gamma}{\sigma}\right)\right].$$

The Birnbaum-Saunders distribution is rarely encountered outside the reliability literature. Several authors who have compared the inverse Gaussian and Birnbaum-Saunders distributions have stated their preference for the inverse Gaussian, which is based on a continuous process instead of an approximation to a discrete one and offers the advantages of belonging to an exponential family whereas the Birnbaum-Saunders distribution does not (see, for example, Bhattacharyya and Fries [BHA 82]).

2.9. Gamma processes

The main topic of this book is the FHT model based on an underlying Wiener process, which should become an important addition to the toolbox of statistical analysis in the biomedical field. Because increments in the Wiener process follow a normal distribution, both positive and negative values are possible. This means that the path followed by the process is not, in general, monotonic in time. This seems reasonable enough when the underlying process expresses a rather general idea of a patient's health, for example. It would be unreasonable to impose monotonicity because even if someone's condition is generally deteriorating, temporary improvements may occur. However, there are situations where it seems desirable for greater realism to require the underlying process to be monotonic. Most of these situations arise in reliability applications. For example, the degradation of a material or the accumulating damage to a structure might be a one-way process. One FHT model that possesses the property of monotonic paths is based on an underlying gamma process. Van Noortwijk, in his review of reliability applications of the gamma process [VAN 09], attributes the first application of the process to modeling deterioration or degradation to Abdel-Hameed [ABD 75].

A gamma process is a continuous process in continuous time with increments following a gamma distribution. Increments in disjoint time intervals are independent. Thus, the structure is like that of the Wiener process, but with a gamma distribution in place of a normal distribution of increments. In general, if $v(t)$ is a non-decreasing, right-continuous, real-valued function for $t \geq 0$ with $v(0) = 0$, then the increment in a time interval $[t_1, t_2]$ follows a gamma distribution with shape $v(t_2) - v(t_1)$ and a scale parameter η that remains the same throughout.

From basic properties of the gamma distribution, the value $Y(t)$ of the process at time t follows the gamma distribution $G(\eta, v(t))$. If the event occurs when the process first reaches a boundary value $\mathfrak{b} > 0$, having started from zero, then the survival function is

$$S(t) = P\left(Y(t) < \mathfrak{b}\right).$$

The distribution of lifetimes is

$$\begin{aligned} F(t) &= P\left(Y(t) > \mathfrak{b}\right) \\ &= P\left(\frac{1}{Y(t)} < \frac{1}{\mathfrak{b}}\right) \\ &= P\left(Y^*(t) < \frac{1}{\mathfrak{b}}\right), \end{aligned}$$

where $Y^*(t)$ follows the inverse gamma distribution. Note that the shape parameter of the gamma distribution involves the time t, a property that Singpurwalla called "quite intriguing" [SIN 95, p.92]. From these functions, the likelihood of a set of data can be written down, involving incomplete gamma functions and other special mathematical functions. Details can be found in, for example, Park and Padgett [PAR 05], who prefer to avoid these complicated expressions and instead use a Birnbaum-Saunders approximation.

2.10. The inverse Gaussian process

Further models can be developed along the lines of the Wiener process and the gamma process by changing the distribution of the increments. One recent contribution along these lines is Ye and Chen's introduction of the inverse Gaussian process [YE 14], in which the increments $Y(t) - Y(s)$ in the process $\{Y(t), t \geq 0\}$ follow the inverse Gaussian distribution with parameters

$$\theta = \Lambda(t) - \Lambda(s), \quad \lambda = \eta\left(\Lambda(t) - \Lambda(s)\right)^2$$

employing the parameterization [2.2], where $\Lambda(t)$ is a monotone increasing function. Taking $\Lambda(0) = Y(0) = 0$ implies that $Y(t)$ follows the inverse Gaussian distribution with parameters $\Lambda(t)$ and $\eta\Lambda(t)^2$ from which it follows easily that the first passage time T_D to a boundary at D is given by the distribution

$$P\left(T_D < t\right) = \Phi\left[\sqrt{\frac{\eta}{D}}\left(\Lambda(t) - D\right)\right] - e^{2\eta\Lambda(t)}\Phi\left[-\sqrt{\frac{\eta}{D}}\left(\Lambda(t) + D\right)\right],$$

(see Wang and Xu [WAN 10]). Note that the existence of this explicit form for the lifetime distribution is one advantage that the inverse Gaussian process possesses in comparison to the gamma process.

Ye and Chen go on to consider inverse Gaussian processes with random effects, as well as the introduction of covariates into its parameters. They suggest that the possibility of introducing random effects in various ways makes the inverse Gaussian process a more flexible model than the gamma process, in which a random effect can only be introduced in one way.

2.11. Degradation and markers

The emphasis of this book is on models for lifetime data based on an underlying process that is not observed. Indeed, in most cases, it is unobservable, because it is conceptualized only in general terms - such as a state of health, or quality of life - and cannot be operationalized into measurable form. This formulation covers almost all applications in the biomedical sciences. In reliability, it is more common for the underlying process to be measurable because, for example, it corresponds to the growth of a crack in a material or the amount of wear on a component of a machine. In these situations, it will be usual for the process to be monitored with measurements made at repeated intervals. The extent of the wear or the size of the crack (in general, the amount of degradation) can be modeled in order to predict the remaining useful lifetime of the machine and plan the schedule of maintenance or replacement. This is clearly a different kind of problem from the lifetime data modeling that is our basic topic, and we will not deal with the very extensive literature that it has generated. Useful references include, but are certainly not limited to Si *et al.* on remaining useful life estimation [SI 11]; Lawless and Crowder's work on a gamma process model with covariates and random effects [LAW 04]; van Noortwijk on the application of the gamma process in maintenance [VAN 09]; Gorjian *et al.*'s review of degradation models in reliability [GOR 10]; various articles in the volume edited by Nikulin *et al.* [NIK 10]; Lu *et al.*'s comparison of failure-time and degradation modeling [LU 96]; degradation modeling in the accelerated testing framework (Meeker *et al.* [MEE 98b], Padgett and Tomlinson [PAD 04], Park and Padgett [PAR 05]); and Pettit and Young's Bayesian analysis of an inverse Gaussian degradation process [PET 99]. Whitmore and Schenkelberg analyze degradation in a Wiener process model with a time-scale transformation [WHI 97].

It should be noted that the majority of the work on degradation modeling uses gamma process models or other models, which, like the gamma process, have the property of monotonicity. However, a significant minority uses the non-monotonic Wiener process. One justification for this is that the reason for imposing monotonicity is less compelling when the degradation process is measured with error. Whitmore analyzed degradation by a Wiener process with normally distributed random measurement errors, independent of the degradation process [WHI 95]; see also Lu and Meeker [LU 93].

Let $W(t)$ represent the stochastic process of interest, which may be an observed process such as measured degradation or an unobserved process, as the case may be. Let Y denote the primary response and $\mathbf{x}$ a baseline covariate vector. Cox has distinguished the following four roles that the process $W(t)$ might play [COX 99].

Failure defining. This is the case that is represented by the first passage of the process to a threshold, with failure occurring when $W(t)$ attains this critical value (the health state falls to zero; the crack grows to a size that causes failure). When the process is the Wiener process, then the inverse Gaussian distribution of failure times is obtained, as seen earlier.

Rate determining. In this case, the process does not directly cause failure. However, the hazard rate at time t depends on $W(t)$ (and not otherwise on t). Models of this kind were discussed by Singpurwalla [SIN 95].

On an equal footing with Y. Cox's example for this category is a medical context in which Y is the survival time and the process $W(t)$ is quality of life, which would need to be taken into account if a treatment offered only a small improvement in survival accompanied by a substantial decline in quality of life. Lee and Whitmore pursue this suggestion, adapting the earlier model of Lee *et al.* [LEE 00] as a joint process for latent health status, quality of life and possibly also markers of health status [LEE 02b].

Intermediate response. $W(t)$ can sometimes be regarded as an intermediate variable, that is, it is both a response to the baseline covariate $\mathbf{x}$ and, together with $\mathbf{x}$, an explanatory variable for the main response Y. Cox notes several scenarios in which this may be useful. One is to use the $W(t)$ process to predict failure. Thus, the information on the state $W(t_c)$ at the censoring time t_c might be used as a predictor of failure time, so that W is acting as a surrogate response variable. Cox shows, under rather specific and strong assumptions, what proportion of the information that has been lost by censoring can be recovered by this procedure [COX 83].

Continuing the discussion regarding W as a surrogate response variable, Prentice points out its value in making a trial or other study more feasible in terms of size or duration: "a rare or distal endpoint is replaced by a frequent or proximate endpoint" [PRE 89, p. 431]. He defines a surrogate response as "a response variable for which a test of the null hypothesis of no relationship to the treatment groups under comparison is also a valid test of the corresponding null hypothesis based on the true endpoint" [PRE 89, p. 432]. His further discussion is in terms of failure rates (hazard functions). Similarly, Jewell and Kalbfleisch, who call $W(t)$ a marker process, also express their models in terms of the hazard function [JEW 96]. Section 4.11 presents a model in which an observable marker process is used in conjunction with a latent health process based on a Wiener process.

3

Model Fitting and Diagnostics

3.1. Introduction

The standard approach to fitting parametric regression models to lifetime data is to use maximum likelihood methods. Given a sample of n independent observations $\{(t_i, \mathbf{x}_i, \delta_i), i = 1, \ldots, n\}$, where unit i with covariates $\mathbf{x}_i$ has lifetime t_i and censoring indicator δ_i, the likelihood is

$$L = \prod_{i=1}^{n} f(t_i|\mathbf{x}_i)^{\delta_i} S(t_i|\mathbf{x}_i)^{1-\delta_i}. \qquad [3.1]$$

Using the relationships between the probability density function, hazard function and survival function, this likelihood can be written in various alternative forms, if desired, as in section 1.3.

It should be emphasized that independent censoring is required for this expression to be valid. This concept is explained in detail in the standard texts on survival analysis. Briefly, the requirement is that "the probability distribution of the residual lifetime for those censored must equal that of those who are not censored" (see Hougaard [HOU 99, p. 15]), among units with the same values of the explanatory variables if there are any. Independent censoring does not necessarily mean that the censoring process is entirely unconnected with the data generating process. For example, "the censoring procedure or rules may depend arbitrarily during the course of the study on previous failure times, on previous censoring times, on random mechanisms external to the study, or on values of covariates included in the model" (see Kalbfleisch and Prentice [KAL 02, p. 195]). The example that is usually given of censoring that is not independent is "if individuals are censored selectively or withdrawn from the study because they appear to be at unusually high (or low) risk of failure compared to others on study with the same covariates" (*loc. cit.*). If the unit were removed from the study because it appeared to be close to failure, then the information that it provides would not be described correctly by $T > t$, but should be $T = t + \delta$ for some small δ.

The likelihood given in equation [3.1] can be maximized analytically in only a few special cases, such as the exponential distribution. In general, iterative numerical methods must be used and, in particular, this is true of the FHT regression based on the Wiener process. Note that the parameter vector in the above equation will include the regression coefficients when some or all of the parameters of the distribution depend on covariates. A regression model of this kind introduces no extra complexity conceptually. However, the optimization problem becomes relatively high-dimensional and practical problems may arise.

3.2. Fitting the FHT regression model by maximum likelihood

In order to fit an FHT regression based on an underlying Wiener process, the density and survival functions in [3.1] take the inverse Gaussian forms

$$f(t|y_0, \mu, \sigma^2) = \frac{y_0}{(2\pi\sigma^2 t^3)^{1/2}} \exp\left[-\frac{(y_0 + \mu\, t)^2}{2\sigma^2\, t}\right]$$

$$S(t|y_0, \mu, \sigma^2) = 1 - \Phi\left[-\frac{(\mu\, t + y_0)}{\sqrt{\sigma^2\, t}}\right] - \exp\left(-2y_0\, \mu/\sigma^2\right) \Phi\left[\frac{(\mu\, t - y_0)}{\sqrt{\sigma^2\, t}}\right],$$

where Φ is the standard normal distribution. Parameters μ and y_0 are substituted by [2.1] and σ^2 is set equal to 1. (Other link functions could be used if desired.) Before discussing programs for solving this general problem, one special case should be mentioned. The inverse Gaussian distribution belongs to the exponential family in the generalized linear models (GLM) framework which allows the mean of the distribution to depend on the linear predictor $\beta'\mathrm{x}$. The inverse Gaussian distribution's mean is equal to $y_0/\,|\mu|$ in the "FHT parameterization" and its shape parameter is y_0^2. Consequently, the FHT regression is a GLM when the drift parameter μ depends on covariates but the starting level parameter y_0 does not. For this purpose, the link function $g(\mu) = \beta'\mathrm{x}$ must be the reciprocal, μ^{-1}. Most books on GLM make minimal reference to the inverse Gaussian distribution. An exception is Hardin and Hilbe who specifically state that their aim is to provide more information so that researchers may recognize the practical usefulness of "the most rarely used model of all traditional GLMs" [HAR 07, p. 105]. They give an example of the modeling of data on length of stay in a hospital. This correspondence with GLMs does not necessarily facilitate fitting the FHT regression model because programs for GLMs would not be expected to deal easily with censored data.

Generalized additive models for location, scale and shape (GAMLSS; see Rigby and Stasinopoulos [RIG 05]) form another class in which parameters of distributions are modeled as functions of covariates. This class allows up to four parameters of a distribution to depend on covariates and extensive software is available (Stasinopoulos and Rigby [STA 07]) in the R language, including an extra module for censored data, *gamlss.cens* [STA 16]. However, the parameterization of the inverse Gaussian distribution that we use in FHT regression does not seem to fit readily into this framework.

A distribution such as this inverse Gaussian model with covariates entering two parameters falls, in principle, within the orbit of Burke and Mackenzie's multi-parameter regression [BUR 16a], mentioned in section 1.4, for which software is available in R (package *mpr*, [BUR 16b]). The program allows covariate dependence on up to three parameters and data can be right censored. However, the inverse Gaussian distribution is not included in the distributions offered by the program.

Many computer programs allow the direct maximization of the likelihood [3.1], usually in the form of minimization of minus the log likelihood, using a general-purpose optimization routine with the user providing the function as input. At the time of writing, there appear to be two programs available that package the procedure to make it easier to use. The first is *stthreg* in Stata (see Xiao *et al.* [XIA 12]) and the second is the R package *threg* (see Xiao *et al.* [XIA 15]). Another R package, *invGauss* [GJE 15], fits the model with randomized drift (section 2.6). These three programs are described in the following sections.

3.3. The *stthreg* package

The Stata *stthreg* package (Xiao *et al.* [XIA 12]) calls the Stata command for maximum likelihood estimation of a user-supplied function: *ml model* (to define the model) and *ml maximize* (to carry out the estimation). It uses the *lf* (linear form) option of the four methods available in *ml maximize* and is based on Newton-Raphson optimization. The package contains four commands, as follows:

– *stthreg*: defines the covariates to be used in μ and y_0 and fits the model. This may be specified as the extended threshold regression cure rate model which is explained below;

– *trhr*: a postestimation command to calculate hazard rates for each level of a categorical covariate in relation to a reference level for a specified scenario (values of the other covariates) at specified times;

– *trpredict*: a postestimation command to estimate μ, y_0, density function f, survival function S and hazard h for a specified scenario and specified time;

– *sttrkm*: to calculate the Kaplan-Meier survival curve for each level of a categorical covariate and overlay the fitted survival curve from an FHT model that includes only this covariate (in μ or y_0 or both). Closeness of the two curves for each level of the covariate indicates a good fit of the regression model to the data.

The documentation includes one example comparing FHT regression and Cox regression on the same small dataset, a larger example illustrating the full set of *stthreg* commands, and a third example illustrating the cure rate model.

The extended model that incorporates a cure rate is a mixture model in which a proportion p of the population is susceptible to failure and the remaining $1 - p$ is not susceptible. As in Balka *et al.* [BAL 09], the survival function becomes

$$1 - p + pS(t|\mu, y_0) = 1 - pF(t|\mu, y_0),$$

where $S = 1 - F$ is the survival function in the susceptible population, and the (improper) density becomes pf where f is the density function of lifetimes among the susceptibles. The proportion p can be modeled (via a logit link function) as depending on covariates. When the cure rate model is being used, then the appropriate changes are made in the output of *trhr*, *trpredict* and *sttrkm*.

This use of the mixture model moves away from the attractive feature of the FHT model based on the Wiener process that long-term survival can emerge as a natural feature without the need to introduce special terms to represent it, as is done in mixture models of the above type. As noted earlier, the probability that the process does not reach the boundary when $\mu > 0$ (the defective inverse Gaussian model) is

$$1 - p_0 = 1 - \exp(-2\mu y_0),$$

and the proportion p_0 can be interpreted as a susceptibility rate. Xiao *et al.* [XIA 12, pp. 276–277] justify the need for the cure rate model by claiming that research populations "often have proportions of susceptible and non-susceptible individuals that do not correspond to p_0 and $1 - p_0$. In this common situation, it is better to let the susceptibility rate be a free parameter that is independently linked to covariates". Similarly, Balka *et al.* [BAL 09, p.171] comment that the "defective inverse Gaussian model does have some initial appeal as a cure rate model, but it is somewhat lacking when it comes to actually estimating the cure rate". Thus, they recommend the mixture model as a more useful alternative for fitting data, although acknowledging the loss of the interpretation of the cure rate as a result of the underlying Wiener process.

If the emergence of a cure rate from the underlying model without the mixture extension is regarded as an advantage of the FHT Wiener-based model, then the obvious course to follow in analyzing data in which it appears that there may be a cured fraction (e.g. from inspection of the simple Kaplan-Meier survival estimates) is to fit the basic model, and, subsequently, also try the mixture version if it seems that there is room for improvement in the fit. In the example presented by Xiao *et al.* [XIA 12], the fitted survival from the basic FHT model appears to be too low (compared to Kaplan-Meier in the plot produced by the *sttrkm* command) for shorter survival times in one of the two groups in the data. This is corrected in the extended cure rate model, with an increase in the estimated proportion of susceptibles from $\hat{p}_0 = 0.219$ to $\hat{p} = 0.244$. The change in minus twice the log likelihood is reported to be significant ($P = 0.045$), although it is not clear that the standard test is appropriate in the context of this mixture model (see, for example, Peng *et al.* [PEN 01]).

3.4. The *threg* package

The R package *threg* (Xiao *et al.* [XIA 15]) follows similar lines to the *stthreg* package. The only major difference is that a cure rate model option is not offered. It is based on the R command *nlm* for optimization using a Newton-Raphson method. Beyond the model fitting, further commands include *hr* and *predict*, providing similar output to the *trhr* and *trpredict* commands of *stthreg*. A *plot* command provides the facility to plot fitted survival, hazard and density functions at the different levels of a covariate for a specified scenario of values of the other covariates. The commands are illustrated on two sets of data.

3.5. The *invGauss* package

Another R package, *invGauss* [GJE 15], fits the specific form of the FHT regression model described in section 2.6 which includes a randomized drift [AAL 08]. An individual's value of the drift parameter is assumed to be drawn from a normal distribution with mean that depends on the individual's covariate values in the usual way and variance τ^2 to be estimated.

Aalen *et al.* [AAL 08, Chapter 10] illustrated the model by analyzing data on survival time after diagnosis in a clinical trial concerning treatment of carcinoma of the oropharynx, taken from Kalbfleisch and Prentice, who used it as an illustration of Cox PH regression [KAL 02]. Four covariates are used in the FHT regression analyses: gender, condition (a 4-point scale from 1=no disability to 4=confined to bed), T-stage (an index from 1 to 4 indicating the size and infiltration of the tumor) and N-stage (an index of lymph node metastasis from 0 to 3). Table 3.1 reproduces in the left-hand panel the results of Aalen *et al.*'s "Model 5", in which the drift does not depend on any covariates and the starting value does not depend on gender.

For comparison, we used the *threg* program to fit the same model without randomized drift. The results are given in the right-hand panel of Table 3.1 and, as far as the regression coefficients are concerned, are clearly very similar to those from the randomized model. Table 3.2 shows AIC values from the two models. Including the random effect in the drift produces a small improvement in the value of the criterion.

	invGauss			*threg*		
Parameter	Estimate	s.e.	z	Estimate	s.e.	z
τ	0.565	0.212				
μ	0.748	0.162	4.61	–0.545	0.086	–6.33
y_0						
Intercept	2.036	0.250	8.13	1.828	0.214	8.56
Condition	–0.685	0.100	6.82	–0.640	0.096	6.65
N-stage	–0.134	0.038	3.52	–0.124	0.037	3.34
T-stage	–0.190	0.050	3.81	–0.182	0.049	3.69

Table 3.1. *Comparison of estimates of the same FHT regression model using the R programs* invGauss *(with randomized drift, variance τ^2) and* threg *(without randomized drift). (Note that the drift is parameterized as $-\mu$ in* invGauss*, but μ in* threg.*)*

	Log likelihood	**AIC**
invGauss	–207.3	426.50
threg	–208.9	427.81

Table 3.2. *AIC values for the fitted FHT regression models in Table 3.1.*

3.6. Fitting FHT regressions using the EM algorithm

A characteristic feature of lifetime data is that the data are usually incomplete in some way. This is because of the presence of right-censored observations. Other possibilities include left censoring, truncation and interval censoring. One general approach to obtaining maximum likelihood estimates from incomplete data is to use the expectation-maximization (EM) algorithm. Indeed, one of the categories of examples given by Dempster *et al.* [DEM 77] in the well-known paper in which they presented the EM algorithm was to these types of incomplete data. An inverse Gaussian regression model is fitted by means of the EM algorithm by Whitmore [WHI 83]. However, this is not the general FHT regression model that is presented in this book. Whitmore uses a parameterization in which a parameter δ corresponding to the reciprocal of the mean is modeled as a linear combination of covariates. This means that he is actually working in the special case of the GLM framework noted above. Balka *et al.* [BAL 09] also use the EM algorithm for inverse Gaussian regression with only the drift parameter depending on covariates, but using the identity link function. Furthermore, the EM algorithm is frequently applied to the category of problems involving mixtures. Balka *et al.* [BAL 09] consequently use it for fitting an inverse Gaussian cure rate model of the kind referred to in section 3.3, with drift and cure rate depending on covariates. They also fit a two-barrier process (see sections 4.6 and 6.12).

3.7. Bayesian methods

A Bayesian approach to an inverse Gaussian regression model was first studied by Upadhyay *et al.* [UPA 96]. The detailed context was a model in which the drift parameter δ depended linearly on a single covariate, $\delta = \alpha + \beta x$, with restrictions on the coefficients to ensure that absorption is certain, because a cured fraction was not envisaged. Vague priors were assumed for the parameters. Two techniques were applied: the Gibbs sampler and sampling-importance resampling.

This line of development was continued by its application to the defective inverse Gaussian regression model by Balka *et al.* [BAL 11], differing from the previous application in that no restrictions were placed on the signs of α, β and $\alpha + \beta x$ in the illustrative application to the case of a single covariate. Vague priors were assumed and the Gibbs sampler was used, together with a chained data augmentation procedure for completing the data.

Bayesian methodology was also used by Pettit and Young, who fitted inverse Gaussian lifetime distributions with additional data available in the form of a measure of the degradation at the time of censoring for those units that had not yet failed [PET 99].

Subsequently, Bayesian methodology was used for the full FHT regression model with both drift and starting level affected by covariates by Pennell *et al.* [PEN 10]. In addition, unmeasured heterogeneity was introduced by including individual random effects in the linear predictors for both μ and y_0. Thus, the regression models were:

$$\mu_i = b_i + \boldsymbol{\beta}'\mathbf{x}_i, \quad \text{with } b_i \sim N(0, \lambda^{-1}),$$

$$y_{0i} \sim N_+(\boldsymbol{\alpha}'\mathbf{u}_i, \tau^{-1}),$$

where N_+ denotes a truncated normal distribution. This analysis will be presented in more detail in section 4.4.

3.8. Checking model fit

There appears to be no test for the suitability of the FHT regression models in the way that specific tests exist to examine the PH (see section 1.5) and AFT (see section 1.7) assumptions. It is necessary to fall back on general methods, applicable to any model. One such method is to compare actual and fitted outcomes, as explained in detail by Lee *et al.* [LEE 10c].

Let the indicator variable $Y_x^{(i)}$ denote the actual outcome for subject i with value x of a covariate and $P_x^{(i)}$ denote the true survival probability given this value. Then, $Y_x^{(i)} - \left(1 - P_x^{(i)}\right)$ has expected value 0 and variance $P_x^{(i)}\left(1 - P_x^{(i)}\right)$. Substitute the estimate $\hat{P}_x^{(i)}$ from the model for the true probability and construct

$$\sum_i \left\{Y_x^{(i)} - \left(1 - \hat{P}_x^{(i)}\right)\right\}, \qquad [3.2]$$

where the summation is over all subjects with the covariate value x. Lee *et al.* claim that it is reasonable to approximate the variance of this sum by the sum of the individual variances, in which case the ratio of the sum [3.2] to its standard deviation is approximately a standard normal deviate, assuming the correctness of the FHT model. This ratio can be plotted against different values of x in order to verify that the pattern appears to be random and unbiased with no outliers. Another test based on the estimated survival probabilities is as follows. Partition the survival times into a set of non-overlapping intervals. The observed numbers of events in each interval can then be compared to the expected numbers under the fitted model using a chi-squared test, where the expected number in an interval is the sum of estimated probabilities for all the subjects in that interval [LEE 02a]. Further mention will be made of model checking in the descriptions of several applications of FHT regression modeling (see, in particular, sections 6.9 and 6.10). Several methods are listed in section 4.11.

It is important to detect outliers - points that are not fitted closely by the model - because we may learn how to improve the model in order to account for all the data satisfactorily. Outlying points may have strong influence (see section 3.10) on the results of modeling, although many survival studies are so large that this is not very likely to be a serious problem. Outliers may be detected informally from plots, as above, or may be supported by formal tests in some cases. Stogiannis and Caroni described some outlier tests for the inverse Gamma distribution, with application to the FHT model [STO 12]. These tests are based on likelihood ratio tests for equality of the inverse Gaussian parameters (θ, λ) in the parameterization of equation [2.2] between units, groups of units or repeated observations on the same unit [DAV 80]. The test for equal values of θ requires equality of the λ but, because of this restriction, is equivalent to a test of equal drift in the FHT framework. However, this is not obviously useful in the regression context. The test for equal λ, on the other hand, is equivalent to a test for equal starting levels in the FHT framework, with no restriction on the drift parameter. Consequently, it could be used when only the drift parameter depends on covariates.

3.9. Issues in fitting inverse Gaussian FHT regression models

3.9.1. *Possible collinearity?*

In the inverse Gaussian FHT regression model, a covariate x_i can affect lifetimes in two ways through the two coefficients β_i and γ_i. The directions of these two effects are in agreement if the coefficients have the same sign. For example, when both β_i and γ_i are positive, individuals with higher values of x_i have a higher initial level as well as a slower drift towards absorption, features which both indicate a tendency to survive longer than individuals with lower values of x_i. On the other hand, opposite signs for β_i and γ_i conflict in the sense that one is indicating a better outcome (longer lifetimes) but the other a poorer outcome (shorter lifetimes). This may sometimes be difficult to interpret and explain.

It is noticeable that "opposite" effects seem to appear quite often in published applications. Examples include results given by Eberly *et al.* [EBE 01], Lee *et al.* [LEE 09], Stogiannis *et al.* [STO 11], and in several other applications that will be cited and discussed in later sections. In Lee *et al.* [LEE 09], the same three predictors of mortality in U.S. railroad workers were included in both μ and y_0: these were indicator variables for smoking, exposure to asbestos and working as an engineer or brakeman (see section 6.3). For lung cancer mortality, each one of the three predictors had a positive coefficient for y_0 (higher starting level) but a negative coefficient for μ (faster drift towards the boundary). In the case of the variable "engineer", the authors suggested that workers attracted into the engineer-brakeman occupation could have been healthier than other workers initially, but subsequently, because of exposure to diesel exhaust fumes, declined more rapidly in health than other workers. However, it is harder to see why workers who smoked or were exposed to asbestos would have had a higher health status initially.

There may sometimes be a good reason to allow a particular covariate x_i to appear in only one of the parameters of the model. For example, the treatment that is received in a randomized trial logically cannot be included in the predictor for y_0, because it comes later in time (see Pennell *et al.* [PEN 10]). Aalen and Gjessing suggested that covariates can be divided into two types, namely those that measure how far the process has advanced and those that represent causal influences on its development. The former should be modeled as influencing y_0 and the latter as influencing μ [AAL 01]. However, many commonly used covariates, such as the age of a patient, could just as well be related to μ as to y_0 in most situations. Therefore, it will usually be necessary in the fitting of the model to allow many of the covariates to affect both of the parameters.

The possibility that a covariate may have similar effects on lifetimes either through its effect on μ or on y_0 led Eberly *et al.* to raise the question of whether these effects can in fact be distinguished in the fitting of the model. In the discussion of Aalen and Gjessing's review, they commented that "it is not clear if identifiability issues arise when covariates are incorporated into both" (see [EBE 01, p.19]). Similarly, Lee and Whitmore remarked in their review that "it will be difficult to attribute the effect to a particular model component ... estimates of covariate effects of the initial value y_0 and mean parameter μ may be collinear" (see [LEE 06, p. 505]).

The main purpose of an investigation by Stogiannis and Caroni described in the following section was to obtain empirical evidence on the possibility of distinguishing the effects of a covariate on lifetimes through its effects on μ and y_0, and whether the issue of potential intrinsic unidentifiability or multicollinearity has practical importance [STO 13]. Is it likely, for example, that the model fitting procedure will place a covariate in the "wrong" parameter? Furthermore, does the observed occurrence of opposite signs of coefficients tend to arise when the covariate ought to have been placed in only one, not both, of the parameters? These questions were addressed in a simulation study.

3.9.2. *Fitting inverse Gaussian FHT regression: a simulation study*

In Stogiannis and Caroni's simulation study [STO 13], data were generated from an inverse Gaussian FHT regression model with either one or two covariates, for selected values of the coefficients that define the covariates' effects on μ and y_0. As pointed out by Eberly *et al.*, the numerical values of these coefficients have no direct interpretation [EBE 01]. Therefore, their values were selected according to the size of the effect on the expected lifetime $E(T) = y_0 / |\mu|$. This makes it possible to choose values that are, in this sense, comparable between the two parameters. If a single covariate x_1 affects y_0 ($\gamma_1 \neq 0$) but not μ ($\beta_1 = 0$), then the effect on the expected lifetime of a unit increase in the value of x_1 is

$$E(T_{x_1+1}) = E(T_{x_1})\, e^{\gamma_1},$$

for any x_1. Values of γ_1 are selected to obtain a desired percentage change in $E(T)$. For example, $\gamma_1 = -0.1054$ gives a 10% decrease in $E(T)$.

If a single covariate affects μ ($\beta_1 \neq 0$) but not y_0 ($\gamma_1 = 0$), the change in $E(T)$ is obtained from

$$E(T_{x_1+1}) = E(T_{x_1}).\, |\beta_0 + \beta_1 x_1| / |\beta_0 + \beta_1(x_1 + 1)|,$$

which depends on x_1. If x_1 is restricted to have range $[0, 1]$ and $x_1 = 0$, then the factor $|\beta_0|/|\beta_0 + \beta_1|$ represents the change from one extreme of the range of values of the covariate ($x_1 = 0$) to the other ($x_1 = 1$). By choosing $\beta_0 = -1$, then, for example, a 10% decrease in $E(T)$ for a unit increase in x_1 is obtained by solving

$$1/|\beta_1 - 1| = 0.9,$$

giving $\beta_1 = -0.1111$ or $\beta_1 = 2.1111$. The first solution is relevant if the aim is to concentrate on the case $\mu < 0$. To investigate the case $\mu > 0$, choosing $\beta_0 = 0.5$, the corresponding equation becomes

$$1/|1 + 2\beta_1| = 0.9,$$

from which we take the larger solution for β_1. Values of γ_0 were chosen so that the probability of not reaching the boundary, evaluated at the mean of x_1, was about 0.5. For each individual in the simulated dataset, the inverse Gaussian parameter values were obtained after the values of the covariate(s) for that individual had been generated. In the event that the parameter $\mu > 0$ for this individual, so that absorption at the boundary was not certain, a pseudorandom uniform variate on $[0, 1]$ was generated first. If this was less than $\exp(-2y_0\mu)$, a lifetime was generated. Otherwise, the lifetime was infinite - in practice, it was set equal to a large censoring time.

Data were generated under three different basic scenarios for the case of a single covariate x_1, generated from the uniform distribution on $[0, 1]$:

a) the covariate affects only y_0 ($\gamma_1 \neq 0, \beta_1 = 0$);

b) it affects only μ ($\gamma_1 = 0, \beta_1 \neq 0$);

c) it affects both y_0 and μ ($\gamma_1 \neq 0, \beta_1 \neq 0$). The cases where γ_1 and β_1 have effects in the same or opposite direction are distinguished.

One way in which the potential "intrinsic collinearity" could affect estimation would be to transfer a covariate's effect to the "wrong" parameter, or to spread the effect across both parameters. Both ways would have an impact on the performance of tests of the null hypotheses. Consequently, the sizes of these tests were checked in the first two scenarios. Overall, the simulated sizes of likelihood ratio and Wald tests for $\gamma_1 = 0$ when $\beta_1 \neq 0$ and for $\beta_1 = 0$ when $\gamma_1 \neq 0$, with $\mu < 0$, were close to the nominal 5% level even when the sample size was only 50.

The results did not depend on the value of the non-zero parameter. In the third scenario, the percentage of 500 runs, in which both the null hypotheses $\beta_1 = 0$ and $\gamma_1 = 0$ were rejected, was also recorded. For example, when the values of β_1 and γ_1 both represent a 75% increase, the simulated powers are 35.2% for rejecting $\gamma_1 = 0$ and 21.6% for rejecting $\beta_1 = 0$. The expected percentage of simultaneous rejections under independence of the two tests is therefore $35.2 \times 21.6/100 = 7.6\%$. However, the observed percentage was only 1.6%. Thus, when the covariate affects both y_0 and μ in the same direction, the two hypotheses $\beta_1 = 0$ and $\gamma_1 = 0$ are rejected simultaneously less often than they would be if the tests of $\beta_1 = 0$ and $\gamma_1 = 0$ were independent. This indicates that there is some tendency for the fitted model to concentrate the two similar effects into one parameter. On the other hand, when the covariate affects the parameters y_0 and μ in opposite directions, the two hypotheses are rejected simultaneously more often than if the tests were independent. These effects can be fairly large. For example, when the values of β_1 and γ_1 both correspond to a 50% reduction in $E(T)$, the frequency of simultaneous rejection of the two hypotheses is 10.4% compared to the 20.1% expected if the two tests were acting independently.

This result seems to indicate that there is some tendency to place two effects with similar directions of action into only one parameter. However, both β_1 and γ_1 are estimated without apparent bias overall.

These simulation results for a single covariate provide some information about the estimation of the coefficients when a covariate appears in the linear predictors of both parameters. Therefore, in the case of two covariates, only the scenario in which the covariates affect different parameters needs to be examined. The first covariate affects μ while the second affects y_0. Consequently, β_1 takes various values in the study but $\beta_2 = 0$ always. Similarly, γ_2 varies but $\gamma_1 = 0$ always. Both covariates may enter both parameters in the fitting of the FHT regression model. As in the previous runs, the simulated sizes of likelihood ratio tests remain acceptably close to the nominal 5% level.

One finding is that the power of the likelihood ratio test for rejecting $\beta_1 = 0$ increases with increasing γ_2 for fixed β_1. In this case there is a tendency for the power for the rejection of $\gamma_2 = 0$ at the nominal 5% level to decline as β_1 increases for fixed γ_2. The explanation of these findings is that larger values of γ_2 imply longer lifetimes. This makes it easier to detect variations in lifetimes that are due to changes in μ in the denominator. Conversely, larger values of β_1 imply shorter expected lifetimes and it becomes relatively more difficult to detect variations that are due to changes in y_0.

Table 3.3 shows, in the left-hand columns, the percentage of times that the estimated values of β_1 and γ_1 have the same sign, as a function of the true value of $\beta_1 \neq 0$ (with $\gamma_1 = 0$). These are average percentages over a range of values of γ_2 between $\pm 75\%$, as there appears to be no dependence on γ_2. The results show a quite strongly increasing tendency for $\hat{\gamma}_1$ to have the same sign as $\hat{\beta}_1$ as the true effect β_1 increases. Similarly, in its right-hand columns, the same table shows the percentage of times that the estimated values of β_2 and γ_2 have the same sign, as a function of the true value of $\gamma_2 \neq 0$ (with $\beta_2 = 0$), again averaged over a range of values of β_1. The increasing tendency for $\hat{\beta}_2$ to have the same sign as $\hat{\gamma}_2$ as the true effect γ_2 increases is evident. In other words, when a covariate has a strong effect on one parameter, there is a tendency for there also to appear to be an effect in the same direction on the other parameter. This tendency increases somewhat as the sample size increases from 50 to 100.

Effect size of β_1	$\hat{\beta}_1, \hat{\gamma}_1$ same sign (%) $n = 50$	$n = 100$	Effect size of γ_2	$\hat{\beta}_2, \hat{\gamma}_2$ same sign (%) $n = 50$	$n = 100$
+75%	38.7	43.3	+75%	44.4	47.7
+50%	32.8	37.7	+50%	38.3	45.3
+25%	28.1	30.4	+25%	26.5	32.9
–25%	28.4	33.0	–25%	32.5	37.8
–50%	42.6	48.7	–50%	48.9	49.8
–75%	52.7	52.0	–75%	49.3	49.7

Table 3.3. *Percentage of 1000 simulation runs in which estimates of β_1 and γ_1, or β_2 and γ_2 have same sign, in relation to true value of β_1 or γ_2, respectively [STO 13]*

The results of the simulation study indicate that the estimation of the regression coefficients is well behaved in certain respects. Even though the sample sizes are relatively small, the sizes of the likelihood ratio and Wald tests are correct and there is no apparent bias in the estimated values of the coefficients. However, the results do confirm that there is some dependence between the estimates of the various coefficients. In particular, when a covariate affects both y_0 and μ in the same direction, there is a tendency for only one of its two coefficients to be identified as statistically significantly different from zero. This did not happen when the covariate affects y_0 and μ in opposite directions; the tendency then is for both to be identified as statistically significantly different from zero.

3.9.3. *Fitting the wrong model*

The preceding results are based on fitting the FHT model to data known to be generated by an FHT model. They allow for model misspecification only to the extent that the fitting procedure allows a covariate to enter both linear predictors whereas it should only be in one. The simulation study of Stogiannis and Caroni also investigates briefly the effect of mistakenly adopting the FHT model when another model is correct [STO 13]. One illustrative example along these lines had been given earlier by Stogiannis *et al.*, who fitted the FHT model to data generated by a Weibull regression, and vice versa [STO 11]. First, PH data with four covariates were generated by a Weibull regression model but the inverse Gaussian FHT model was fitted with all four covariates allowed to enter the linear predictors for both μ and y_0. Within just this one analysis, a variety of outcomes were produced: one covariate was statistically significant only in the linear predictor for μ, two were significant only for y_0, and one was significant for both μ and y_0. Second, data were generated from the inverse Gaussian distribution using the same data generation scheme for the four covariates, but with two of them affecting only μ and the other two affecting only y_0. All the coefficients were negative, meaning that higher values of the covariates would tend to be associated with shorter lifetimes. Consequently, in the Cox regression fitted to these data, positive coefficients (increasing hazard) would be expected. In fact, although the sample size was large ($n = 100$), only the two covariates that affected μ were statistically significant and one of them had a negative coefficient. These illustrative results suggest that PH and inverse Gaussian FHT regression models are liable to give quite different results when fitted to the same data, and hence the choice of the appropriate model is a matter of great importance.

To investigate the effect of model misspecification further, Stogiannis and Caroni repeatedly fitted the inverse Gaussian FHT regression model to sets of simulated data that had been generated by a Weibull regression with two covariates [STO 13], the first generated from the uniform distribution U[0,1] and the second independently from N(2,1). About 30% censoring is introduced by generating an independent censoring time from an exponential distribution. Two values of each Weibull regression coefficient are used in the simulations chosen, so that the power for rejecting the null hypothesis of zero coefficient when the Weibull was fitted to these data was about 75% and 95% for the two values, respectively. The results of fitting the inverse Gaussian FHT regression model to these data are shown in Table 3.4 for 500 simulation runs for each combination of true values of the Weibull regression coefficients. The principal observations are that: (a) the FHT fitting is about twice as likely to place the effect of a covariate in y_0 as in μ, (b) both coefficients for the same hypothesis are rejected simultaneously less often than if the tests for the two coefficients were independent and (c) when both are rejected, the coefficients tend to have opposite signs.

	True values of Weibull regression parameters			
	$(1.1, 0.3)$	$(1.1, 0.5)$	$(1.5, 0.3)$	$(1.5, 0.5)$
Reject $\beta_1 = 0\,(\%)$	22.9	24.9	29.2	29.3
Reject $\gamma_1 = 0\,(\%)$	59.0	53.6	58.1	62.9
Reject both				
Expected (%)	13.5	13.3	17.0	18.4
Observed (%)	11.2	10.9	10.3	11.3
With the same sign (%)	0.6	0.3	1.6	3.0
Reject $\beta_2 = 0\,(\%)$	23.0	29.1	27.1	34.2
Reject $\gamma_2 = 0\,(\%)$	52.4	59.6	54.4	61.5
Reject both				
Expected (%)	12.1	17.3	14.7	21.0
Observed (%)	8.8	9.7	11.4	11.7
With the same sign (%)	0.4	4.4	1.3	5.3

Table 3.4. *Results of likelihood ratio tests of coefficients in the FHT model fitted to data generated from Weibull regression with two covariates, $n = 100$ (adapted from [STO 13])*

The last of these conclusions suggests that the finding of opposite directions of the two effects of the same covariate – which seems to be common in published applications – might tend to arise through trying to fit the FHT model when it is not the correct one. This in turn indicates the importance of developing diagnostics and tests of goodness of fit for the FHT model.

3.10. Influence diagnostics for an inverse Gaussian FHT regression model

Model fitting should be followed by the application of diagnostic methods in order to assess goodness of fit and influence. As noted in section 3.8, diagnostics aimed specifically at the FHT regression model have yet to be developed in the literature, although it will be seen in the applications given in later sections that various general ideas have been used. Diagnostics specifically developed for the inverse Gaussian FHT regression model should take account of the distinctive feature of the model, namely, that dependence on a particular covariate can appear in either y_0 or μ, or both.

We construct here a case-deletion diagnostics method (CDM) for the FHT regression model based on the general approach introduced by Cook and Weisberg [COO 82]. The CDM is a technique of major importance because of the ease of its applicability. A way to measure the influence of the ith observation on the parameter estimates is to compute single-case diagnostics with the ith case deleted. We use the same idea to modify the measure in such a way as to focus interest on only one of the two parameters of the inverse Gaussian distribution.

The log likelihood displacement or distance LD_i is defined for the ith of n cases as

$$LD_i = 2\left[\ell(\hat{\boldsymbol{\theta}}) - \ell(\hat{\boldsymbol{\theta}}_{(i)})\right], \qquad [3.3]$$

where $\hat{\boldsymbol{\theta}}_{(i)}$ is the maximum likelihood estimate of $\boldsymbol{\theta}$ obtained from the reduced dataset of $n-1$ observations after omitting the ith observation. LD_i is a measure of the change in the parameter estimate from $\hat{\boldsymbol{\theta}}$ to $\hat{\boldsymbol{\theta}}_{(i)}$ as a result of omitting this observation. This is a special case of data perturbation via an n-vector of weights ω such that

$$\ell(\theta|\boldsymbol{\omega}) = \sum \omega_i \ell_i,$$

where ℓ_i is the contribution made by the ith case to the log likelihood. For unperturbed data, $\boldsymbol{\omega} = (1, 1, \ldots, 1)'$. Deletion of the ith case corresponds to the choice $\boldsymbol{\omega} = (1, \ldots, 1, 0, 1, \ldots, 1)'$, where the zero appears in the ith position and all other elements equal one. Although the need for approximations to ease the computational burden is unimportant these days, we observe that the expression given in equation [3.3] can be approximated using a Taylor expansion of $\ell(\hat{\boldsymbol{\theta}}_{(i)})$ around $\hat{\boldsymbol{\theta}}$.

$$\ell(\hat{\boldsymbol{\theta}}_{(i)}) \approx \ell(\hat{\boldsymbol{\theta}}) + \left(\hat{\boldsymbol{\theta}}_{(i)} - \hat{\boldsymbol{\theta}}\right)' \dot{\boldsymbol{\ell}}\left(\hat{\boldsymbol{\theta}}\right) + \frac{1}{2}\left(\hat{\boldsymbol{\theta}}_{(i)} - \hat{\boldsymbol{\theta}}\right)' \ddot{\boldsymbol{\ell}}\left(\hat{\boldsymbol{\theta}}\right)\left(\hat{\boldsymbol{\theta}}_{(i)} - \hat{\boldsymbol{\theta}}\right)$$

and, since $\dot{\boldsymbol{\ell}}\left(\hat{\boldsymbol{\theta}}\right) = \mathbf{0}$, this gives

$$LD_i \approx -\left(\hat{\boldsymbol{\theta}}_{(i)} - \hat{\boldsymbol{\theta}}\right)' \ddot{\boldsymbol{\ell}}\left(\hat{\boldsymbol{\theta}}\right)\left(\hat{\boldsymbol{\theta}}_{(i)} - \hat{\boldsymbol{\theta}}\right).$$

Thus, the log likelihood distance is approximately a form of the generalized Cook's distance presented by Cook and Weisberg [COO 82]

$$GD_i = \left(\hat{\boldsymbol{\theta}}_{(i)} - \hat{\boldsymbol{\theta}}\right)' \mathrm{M}\left(\hat{\boldsymbol{\theta}}_{(i)} - \hat{\boldsymbol{\theta}}\right),$$

where $\mathbf{M}$ is a non-negative definite matrix. We can see how to adapt this to the case in which a subset of the parameters is of interest by considering the asymptotic confidence region for $\boldsymbol{\theta}$

$$\left\{\boldsymbol{\theta} : 2\left[\ell(\hat{\boldsymbol{\theta}}) - \ell(\boldsymbol{\theta})\right] \leq \chi^2_{q\,;\alpha}\right\},$$

where $\chi^2_{q\,;\alpha}$ is the upper $\alpha\%$ point of the χ^2 distribution with degrees of freedom q given by the dimension of $\boldsymbol{\theta}$.

If the subset $\boldsymbol{\theta}_1$ is the object of interest, an asymptotic confidence region for it is:

$$\left\{\boldsymbol{\theta}_1 : 2\left[\ell(\hat{\boldsymbol{\theta}}) - \ell\left(\boldsymbol{\theta}_1, \boldsymbol{\theta}_2(\boldsymbol{\theta}_1)\right)\right] \leq \chi^2_{q_1\,;\alpha}\right\},$$

(see Cox and Hinkley [COX 74]), where $\boldsymbol{\theta}' = (\boldsymbol{\theta}_1', \boldsymbol{\theta}_2')$, q_1 is the dimension of $\boldsymbol{\theta}_1$ and

$$\ell\left(\boldsymbol{\theta}_1, \boldsymbol{\theta}_2(\boldsymbol{\theta}_1)\right) = \max_{\boldsymbol{\theta}_2}\left[\ell(\boldsymbol{\theta}_1, \boldsymbol{\theta}_2)\right].$$

This suggests defining the measure of the displacement of $\hat{\boldsymbol{\theta}}_1$ to $\hat{\boldsymbol{\theta}}_{1(i)}$, when the ith case is deleted, as the profile log likelihood displacement [ESC 92]:

$$\begin{aligned} LD_i(\boldsymbol{\theta}_1|\boldsymbol{\theta}_2) &= 2\left[\ell(\hat{\boldsymbol{\theta}}) - \ell\left(\hat{\boldsymbol{\theta}}_{1(i)}, \boldsymbol{\theta}_2(\hat{\boldsymbol{\theta}}_{1(i)})\right)\right] \\ &= 2\left[\ell(\hat{\boldsymbol{\theta}}) - \max_{\boldsymbol{\theta}_2} \ell(\hat{\boldsymbol{\theta}}_{1(i)}, \boldsymbol{\theta}_2)\right]. \end{aligned}$$

In the FHT regression context, the parameter vector is $\boldsymbol{\theta}' = \left(\boldsymbol{\beta}', \boldsymbol{\gamma}'\right)$ as in equation [2.1] and the role of $\boldsymbol{\theta}_1$ is taken by either $\boldsymbol{\beta}$ or $\boldsymbol{\gamma}$ in order to examine influence on the estimate of the mean or starting level, respectively. Another choice that may be interesting is $\boldsymbol{\theta}_1' = (\beta_i, \gamma_i)$, which examines the simultaneous influence on the estimates of both of the coefficients related to the covariate x_i and hence examines the overall influence with respect to that covariate.

3.11. Variable selection

In many lifetime regression analyses in the biomedical sciences, a large number of potential regressors are available (e.g. recorded in the patient's file).

Part of the object of the analysis is to identify which of these need to be retained in the final regression model. This issue is much less likely to arise in reliability analysis, where covariates tend to be few (possibly because of the difficulty or expense of measurement, or the unfeasibility of running an accelerated life test with many acceleration factors) and have often been selected on the basis of their known association with the outcome based on previous experience.

Variable selection procedures have been developed extensively in the context of other regression models, notably for the general linear model. Earlier methods were stepwise procedures, such as the successive elimination of coefficients that fail to meet a prespecified level of statistical significance, followed by refitting the model and repeating until no more covariates can be eliminated. These procedures can be automated, as implemented in some well-known statistical packages, or carried out manually. A procedure along these lines that can be found in many investigations is first to determine which of the potential regressors has an association with the outcome in a bivariate analysis, then second to conduct a regression analysis that includes only the regressors retained from the first stage and third to refine this regression model by eliminating non-significant regressors. Wald or likelihood ratio tests would likely be used for the decisions on significance. In Cox regression, the score test is also used because of its equivalence to the well-known log rank test. However, stepwise procedures do not have optimal properties.

More modern approaches include procedures that simultaneously execute estimation and variable selection, notably the lasso method introduced for the general linear model by Tibshirani [TIB 96] and developments of it. Tibshirani subsequently applied the method to Cox regression [TIB 97]. The methods for the Cox model are available in various R packages including *penalized* [GOE 17].

The FHT regression model based on the Wiener process differs from the context of these methods in that estimation and selection must be applied to two vectors of regression coefficients. Stogiannis and Caroni [STO 16] have considered applying Wang and Leng's least squares lasso approximation [WAN 07] to each of the vectors separately. Burke and MacKenzie approach variable selection in their multi-parameter regression framework (see section 1.4) in the spirit of stepwise procedures, although taking into consideration the need to test for the simultaneous exclusion of the two coefficients related to the same covariate. This is included in their *mpr* package (see section 3.2).

As already mentioned, this package does not include the inverse Gaussian distribution as an option. A lasso method applied to inverse Gaussian regression is presented by Wu and Li [WU 12]. As given, this operates in a parameterization similar to that of equation [2.2], not the FHT regression parameterization, but the method should be adaptable to the latter.

4

Extensions to Inverse Gaussian First Hitting Time Regression Models

4.1. Introduction

The vast majority of the countless applications of Cox's semi-parametric Proportional Hazards (PH) regression model use the basic model described earlier in section 1.14. However, that has been extended in various ways, in order to meet the demand for models that can handle time-varying covariates and recurrent events, for example. Several of these extensions are described in detail by Therneau and Grambsch [THE 00] and others. Equivalent extensions and more can be found among recent developments of the inverse Gaussian First Hitting Time (FHT) regression model and are described in this chapter.

4.2. Time-varying covariates

The basic idea of a regression model for a lifetime T is to introduce in some way the lifetime's dependence on a vector of covariates $\mathbf{x}$. The values of the elements of $\mathbf{x}$ are recorded when the unit enters the study at time zero. Some of these covariates have values that, by their nature, never change. Obvious examples include the batch in which a mechanical part was manufactured, a patient's gender, and the type of surgery that was carried out. Other covariates do change and it seems reasonable to suppose that, since we are accepting that lifetime may depend on these covariates, our models will be improved by taking into account updated values instead of relying on the values of the covariates that were recorded at time zero. In most cases, the availability of updated values on these time-varying covariates implies that a longitudinal study is being carried out.

Two principal types of time-varying covariates are distinguished in the literature, *external* and *internal* (see Kalbfleisch and Prentice [KAL 02]). External covariates may be further separated into *defined* and *ancillary*. The standard example of a defined covariate is age, which follows a determined time path. An ancillary covariate could be an environmental factor (weather, air pollution) at the location where the study is being carried out. In contrast to these examples, an internal covariate is related to the individual's behavior. Smoking is one example: an individual's smoking habits may change over the duration of a study. A marker, in the meaning of section 2.11, is also an internal time-dependent covariate.

It is important to be aware that the use of time-varying covariates, especially internal ones, is full of dangers. For example, a change in smoking behavior may be due to changes in the individual's health, so it has the character of an outcome variable rather than a predictor. Furthermore, the nature of the model changes. PH models are no longer PH and well-known relationships such as that between hazard and survival no longer hold. Many of these issues are discussed in the context of Cox's PH model by Fisher and Lin [FIS 99].

A standard way of carrying out the computing necessary to fit the Cox PH model with time-varying covariates is as follows. Suppose that data on an individual in a longitudinal study have been recorded at times $0 = t_0 < t_1 < t_2 < \ldots < t_m$, with corresponding covariate vectors $\mathbf{x}_i$ $(i = 0, \ldots, m-1)$ and status indicators δ_i $(i = 1, \ldots, m)$. We restructure or decompose the data file so that this individual appears not as one record but as m records. The ith record has a start time t_{i-1}, finish time t_i, covariate $\mathbf{x}_{i-1}$ and status indicator δ_i (which is $\delta_i = 0$, right-censored, for every record except possibly the last one for the individual). The standard Cox model is then fitted to the restructured data file. This decomposition is carried out automatically by some software, for example by the command *stsplit* in Stata.

The extension to time-varying covariates of the FHT regression model based on an underlying Wiener process is examined in detail by Lee *et al.* [LEE 10b]. They adopt the approach of decomposing the longitudinal record, as above, and investigate the conditions necessary for its validity in the context of the FHT regression model. The key requirement is the assumption of a Markov process. At time t_j $(j = 0, 1, \ldots, m)$ of the longitudinal process, let $\mathbf{x}_j$ be the covariate vector as before, y_j the process level (supposing for the time being that this can be observed), f_j the failure indicator (1 for failure at or before t_j, 0 otherwise) and δ_j the censoring indicator (1 for censoring at t_j, 0 otherwise). Note that y_m at the final data point may be undefined if $f_m = 1$. Then, the data are $A_j, j = 0, 1, \ldots, m$ with $A_j = (y_j, \mathbf{x}_j, f_j, \delta_j)$ and the probability of the observed sequence is

$$P(A_m, A_{m-1}, \ldots, A_0) = P(A_0) \prod_{j=1}^{m} P(A_j | A_{j-1}, \ldots, A_0).$$

Assuming a Markov property implies that the probability of A_j conditional on the history of the sequence actually depends only on the immediately previous state A_{j-1}, so the above expression simplifies to

$$P(A_m, A_{m-1}, \ldots, A_1) = P(A_0) \prod_{j=1}^{m} P(A_j | A_{j-1}).$$

It is this decomposition into single-step transitions that effectively removes the longitudinal nature of the data, which can now be handled as simply as non-longitudinal data. Furthermore, because $f_j = \delta_j = 0$ for $j < m$, the transition probabilities are:

$$\begin{aligned} &P\left(y_j, f_j | \mathbf{x}_j, \delta_j, y_{j-1}, \mathbf{x}_{j-1}, f_{j-1} = 0, \delta_{j-1} = 0\right) \\ &\times P\left(\mathbf{x}_j, \delta_j | y_{j-1}, \mathbf{x}_{j-1}, f_{j-1} = 0, \delta_{j-1} = 0\right). \end{aligned}$$

The second term in this factorization describes the joint process for the covariate vector and the censoring mechanism. Often, this does not depend on the parameters θ that are involved in the first term. In that case, the partial likelihood that contains only these terms can be used for inference about θ:

$$L(\theta) = \prod_{j=1}^{m} P\left(y_j, f_j | \mathbf{x}_j, \delta_j, y_{j-1}, \mathbf{x}_{j-1}, f_{j-1} = 0, \delta_{j-1} = 0\right)$$

and in addition the δ_j can be dropped if censoring is independent of the process $Y(t)$. Finally, if $Y(t)$ is unobservable - probably the case of greatest practical importance - the terms in the partial likelihood take the form

$$P\left(f_j | \mathbf{x}_j, \mathbf{x}_{j-1}, f_{j-1} = 0\right).$$

Lee *et al.* go on to consider special cases [LEE 10b].

4.3. Recurrent events

In many circumstances, a lifetime is the time until a unique event occurs - obviously so in the case of an individual's death or the catastrophic failure of a machine without possibility of repair. But in other cases, the event could potentially recur, for example, a machine may be repaired and returned to service, or a motorist may make multiple claims against an insurance policy. However, even in these situations, the interest may lie principally in the first event, possibly because subsequent events are thought likely to be governed by different processes, or because the time frame of the study is too short for most individuals to have more than one event. There are, however, situations in which the analysis of the recurrence of events is an important feature of the study. One example is in the study of chronic diseases, because the patient is bound to suffer repeated episodes of illness and it would be almost meaningless to look only at the next occurrence within the study.

An analysis of recurrent events in the context of an underlying Wiener process model is presented by Whitmore *et al.* [WHI 12a], who study the renewal process generated by the first hitting times of the path of the process on successive equally spaced levels $\omega, 2\omega, \ldots$ starting from zero. Four different observation schemes are examined, formed by the combinations of two types of data and two choices of starting time. The starting time can be either a renewal or a random point in a stationary process. The data consist of either the sequence of gap times between successive events within a window of time or simply the count of the number of events within that window. Likelihood estimation methods are provided for each scheme and the relative precisions of the estimates between schemes are examined in a simulation study. For both choices of starting point, gap time data provide more precise estimates of parameters than count data. For the gap time data, observation windows starting with a renewal point provide more precise estimates than the windows starting randomly within a stationary process, but for the count data the opposite is true. The results are used to provide a means of planning a recurrent event study. However, the main interest for present purposes lies in the introduction of covariates. Covariates that are fixed for each individual throughout the observation window can be introduced in the usual way, through their effect on the drift and level parameters of the underlying Wiener process. Furthermore, covariates that are specific to each gap can be included in the same way when the data consist of gap times. The introduction of covariates when the data consist of counts is not pursued in the article.

The proposed methods are applied to a set of gap times between exacerbations experienced by 449 patients suffering from chronic obstructive pulmonary disease (COPD) during a clinical trial. Eight baseline covariates were included: these remained the same for the patient throughout the trial. One gap-specific covariate was used, namely, the sequence number of the gap within the observation window. This was included only in the regression for the drift parameter. The analysis used results for gap times recorded in a window starting from a random point in a stationary process. An earlier analysis of the same data by Aaron *et al.* [AAR 10] had employed, in effect, the analysis for a window starting with a renewal point (see section 6.7). The basic results of the re-analysis showed that the level parameter ω of the Wiener process did not depend on covariates. The rate of the Wiener process depended on several variables including the sequence number of the gap, with a positive coefficient that indicates an acceleration of the process of exacerbations as time advanced.

4.4. Individual random effects

Regression modeling in its basic form depends on an assumption that subjects with the same values of the covariates have the same lifetime distribution. However, at least as far as human subjects are concerned, it is unlikely that the observed covariates account for all the differences between subjects. In addition, unexplained heterogeneity between subjects can be represented by subject-specific coefficients, as in the idea of frailty (section 1.18). Pennell *et al.* aim to introduce such heterogeneity into the starting point and drift of a Wiener process that underlies lifetimes [PEN 10]. They mention earlier work which introduced between-subject heterogeneity only into the drift terms [WHI 86b, AAL 01].

Following Aalen and Gjessing [AAL 01], and also Saebø *et al.* [SAE 05], the usual regression model for the drift parameter μ_i of subject i is augmented by a subject-specific random term b_i:

$$\mu_i = b_i + \beta' \mathbf{x}_i, \quad \text{where} \quad b_i \sim N(0, \lambda^{-1}).$$

The analysis is carried out within a Bayesian framework. Conjugate prior distributions are assigned to the regression parameters β and to the precision parameter λ:

$$\beta \sim N(\beta_{prior}, \Sigma_\beta)$$
$$\lambda \sim Ga(\gamma_1, \gamma_2),$$

where $Ga(\gamma_1, \gamma_2)$ denotes a gamma distribution with mean γ_1/γ_2 and variance γ_1/γ_2^2. While previous analyses might guide the choice of β_{prior} and its variance Σ_β, information is unlikely to exist for λ so a diffuse prior such as $Ga(2, 1/2)$ is suggested.

Extending the regression equation for the starting value in a similar way is a possibility, but not fruitful because conjugate priors do not exist. Computationally, a much more useful device is to assume that the starting values are drawn from a normal distribution truncated on the left at zero (denoted N_+):

$$y_{0i} \sim N_+(\alpha' \mathbf{u}_i, \tau^{-1}).$$

The moments of this distribution are

$$E(y_{0i}|\alpha, \tau) = \alpha' \mathbf{u}_i + \frac{h(\zeta_i)}{\sqrt{\tau}}$$
$$V(y_{0i}|\alpha, \tau) = \tau^{-1}\left\{1 - h(\zeta_i)\left[h(\zeta_i) - \zeta_i\right]\right\},$$

where $\zeta_i = -\tau^{1/2} \cdot \alpha' \mathbf{u}_i$ and $h(.)$ is the hazard function of the standard normal distribution (see Barr and Sherrill [BAR 99]). A normal prior distribution is assumed for the coefficients α

$$\alpha \sim N(\alpha_{prior}, \Sigma_\alpha)$$

and, similarly to the precision parameter λ above, a Gamma distribution could be taken for τ. However, it is claimed that simulation studies indicate that little information is available about this parameter in the data, and therefore it is preferable to fix its value in order to avoid numerical problems. The analysis can be repeated with different values as a sensitivity analysis.

It is now necessary to write down the likelihood function. This differs from the usual form given in equation [3.1] in respect of the right-censored survival times, which are generally represented by terms $S(t)$ where S is the survival function of the inverse Gaussian distribution. This was done by Saebø *et al.* [SAE 05] but it results in complicated posterior distributions. More tractable results are obtained by replacing $S(t)$ by the pdf of the current level y_i of the Wiener process, which is

$$f_0\left(Y_i(t_i) = y_i, t_i < s_i | y_{0i}, \mu_i, \sigma^2\right)$$
$$= \frac{1}{\sqrt{2\pi\sigma^2 t_i}} \exp\left\{-\frac{(y_i - y_{0i} - \mu_i t_i)^2}{2\sigma^2 t_i}\right\}\left[1 - \exp\left(\frac{-2y_i y_{0i}}{\sigma^2 t_i}\right)\right] I(y_i > 0)$$

(see Cox and Miller [COX 65]). Since y_i is unobserved, it will be necessary to treat it as an unknown parameter (to which a flat prior distribution will be assigned, given that it is unlikely that prior information is available). Finally, the likelihood function for the model is

$$L(\mathbf{y}_s, \mathbf{y}_0, \boldsymbol{\beta}, \mathbf{b}) = \prod_{i=1}^{n} \left[f\left(s_i | y_{0i}, \mu_i, \sigma^2\right)\right]^{\delta_i} \left[f_0\left(y_i, t_i < s_i | y_{0i}, \mu_i, \sigma^2\right)\right]^{1-\delta_i},$$

where the vector $\mathbf{y}_s$ contains the unobserved levels at the censoring time for the censored observations and δ_i is the usual censoring indicator.

The estimation of the model involves the application of a data augmentation approach along with a hybrid MCMC algorithm which incorporates both Gibbs sampling and rejection sampling. In each MCMC iteration step, the latter is used to update the levels y_i at censoring times and the former to update the other parameters of the model. The method is illustrated by application to the well-known set of data on survival of patients with malignant melanoma, which previous analyses using frailty models have found to contain unexplained heterogeneity. In comparison to the standard FHT regression model, the inclusion of random effects changed the statistical significance of some covariates but had less effect on predicted survival. However, in an analysis of simulated data generated under the random effects model, survival estimates were quite different between the fitted models with and without random effects.

4.5. First hitting time regression model for recurrent events with random effects

The model introduced by Pennell *et al.* [PEN 10] that was described in the previous section is adapted and extended to recurrent events by Economou *et al.* [ECO 15]. It is supposed that the underlying process for an individual restarts after the occurrence of an event. Thus, the overall life history consists of a number of stages, each one ending in an event or - in the case of the last stage - a right-censored observation time.

The structure of the data for the ith individual is therefore as follows:

Number of observed stages:	n_i
Time points:	$0 = t_{0i} < t_{1i} < t_{2i} < \ldots < t_{n_i i}$
Failure code for the final stage:	$\delta_i = 0$ (censored) or 1(failed)
Covariate vectors:	$\mathbf{x}_{1i}, \ldots, \mathbf{x}_{n_i i}$

where $\mathbf{x}_{mi} = (1, x_{m1\,i}, \ldots, x_{mp\,i})'$, $1 \leq m \leq n_i$ is the vector of the values of the p covariates of the ith individual at the mth stage. These are measured only at the beginning of each stage. In fact, in some applications, the covariates are available only at the beginning of the study, in which case the covariates are constant over time and $\mathbf{x}_{mi} = \mathbf{x}_{1i}$, for all $1 \leq m \leq n_i$.

The regression model with individual random effects for the first stage follows Pennell *et al.* [PEN 10] as in section 4.4:

$$y_{1i} \sim N_+(\boldsymbol{\alpha}'\mathbf{u}_{1i}, \tau^{-1})$$

$$\mu_{1i} = b_{1i} + \boldsymbol{\beta}'\mathbf{v}_{1i}, \quad \text{with } b_{1i} \sim N(0, \lambda^{-1}), \qquad [4.1]$$

where $N_+()$ denotes a normal distribution truncated to the left at zero and $\mathbf{u}_{1i}$ and $\mathbf{v}_{1i}$ are the covariate vectors associated with the initial value y_{1i} and the drift μ_{1i}, respectively, of individual i in the first stage of the process. The sets $\mathbf{u}_{1i}$ and $\mathbf{v}_{1i}$ are subsets of the available covariates $\mathbf{x}$ and not necessarily disjoint.

To represent the relationship between stages of the same individual's life history, the observed covariate vectors can be extended by adding elements that record how many previous events have occurred to that individual and the total time under observation until the beginning of each stage (sum of the previous inter-event times $s_1, \ldots, s_{m-1}$, that is, t_{m-1}). Both of these covariates are allowed to affect the starting level y_{mi} and the drift parameter μ_{mi} for the stages of individual i after the first. Furthermore, we can suppose that the changes to the starting level and the drift parameter of a new stage are related to the starting level and the drift parameter, respectively, of the previous stage, as well as to changes in the values of the covariates between the stages.

The drift parameter in stage i therefore satisfies the following relationship:

$$\mu_{mi} - \mu_{(m-1)i} = b_{mi} + \boldsymbol{\beta}'_d(\mathbf{v}_{mi} - \mathbf{v}_{(m-1)i}) + \beta_{ns} + \beta_s s_{(m-1)i}, \quad [4.2]$$

where $b_{mi} \sim N(0, \lambda^{-1})$ are the individual random effects that represent unexplained between-subject heterogeneity between the drift in each stage and $\boldsymbol{\beta}$ are the regression coefficients. The coefficients with subscripts ns and s introduce the dependence on the number of previous stages and the time elapsed, respectively. The coefficients with subscripts d express the influence of the difference between the values of the covariates of the mth stage and the initial stage.

For a more detailed exposition of this structure, assume that there are two covariates w and y. The former is constant over time (e.g. gender) whereas y may take a different value for each stage. Equation [4.1] gives

$$\mu_{1i} = b_{1i} + \beta_0 + \beta_w w + \beta_y y_{1i}, \quad \text{with } b_{1i} \sim N(0, \lambda^{-1}) \quad [4.3]$$

and [4.2] gives

$$\mu_{mi} - \mu_{(m-1)i} = b_{mi} + \beta_{dy}(y_{mi} - y_{(m-1)i}) + \beta_{ns} + \beta_s s_{(m-1)i}.$$

Consequently, μ_{mi} can be expressed as

$$\mu_{mi} = \sum_{k=1}^{m} b_{ki} + \beta_0 + \beta_w w + \beta_y y_{1i} + \beta_{dy}(y_{mi} - y_{1i}) + \beta_{ns}(m-1) + \beta_s t_{(m-1)i}$$

and, as the random effects b_{ki} are normally distributed, it follows that μ_{mi} is also normally distributed with location parameter

$$\beta_0 + \beta_w w + \beta_y y_{1i} + \beta_{dy}(y_{mi} - y_{1i}) + \beta_{ns}(m-1) + \beta_s t_{(m-1)i}$$

and variance $m\lambda^{-1}$. The correlations between the drift parameters are given by

$$corr(\mu_{mi}, \mu_{(m+r)i}) = \frac{m}{\sqrt{m}\sqrt{m+r}}$$

and the $\mu_{m1}, \mu_{m2}, \ldots, \mu_{mn_m}$ are multivariate normally distributed. This correlation decreases as the difference r increases, expressing the weak correlation between the drifts of well-separated stages. However, the correlation $corr(\mu_{mi}, \mu_{(m+1)i})$ between the drifts of successive stages tends to one as m increases. This follows from the fact that the effect of the cumulative random effect $(\sum_{k=1}^{m} b_{ki})$ becomes more significant as the number of stages increases.

Now consider the starting points of every stage. By following steps similar to those for the starting points, the location parameter of y_{mi} is given by

$$\alpha_0 + \alpha_w w + \alpha_y y_{1i} + \alpha_{dy}(y_{mi} - y_{1i}) + \alpha_{ns}(m-1) + \alpha_s t_{(m-1)i},$$

where $(\alpha_0, \alpha_w, \alpha_y, \alpha_{dy}, \alpha_{ns}, \alpha_s)$ are the regression coefficients. The distribution of y_{mi} is assumed to be truncated normal as before, with this location parameter and dispersion parameter τ^{-1}.

To summarize, in this model the location parameter of both the starting level and the drift can depend on the initial values of the covariates, the changes in values of the covariates from the first stage, the number of previous stages, and the total time on observation until the beginning of each stage. Additionally, the drift depends on all the previous random effects of the individual.

The regression model can be expressed by the following general relationships for every $m \quad (1 \leq m \leq n_i)$:

$$y_{mi} \sim N_+ \left(\boldsymbol{\alpha}^{*\prime}\tilde{\mathbf{u}}_{mi}, \tau^{-1}\right)$$

$$\mu_{mi} = \sum_{k=1}^{m} b_{ki} + \boldsymbol{\beta}^{*\prime}\tilde{\mathbf{v}}_{mi} \qquad [4.4]$$

where $\tilde{\mathbf{u}}_{mi}$ and $\tilde{\mathbf{v}}_{mi}$ are the extended covariate vectors of the mth stage of individual i associated with the initial value y_{mi} and the drift μ_{mi} of the mth stage of the process, respectively. The coefficients $\boldsymbol{\alpha}^{*\prime}$ and $\boldsymbol{\beta}^{*\prime}$ are the regression parameters given by $\boldsymbol{\alpha}^{*\prime} = (\boldsymbol{\alpha}', \boldsymbol{\alpha}'_d, \alpha_{ns}, \alpha_s)$ and $\boldsymbol{\beta}^{*\prime} = (\boldsymbol{\beta}', \boldsymbol{\beta}'_d, \beta_{ns}, \beta_s)$ respectively and so

$$\begin{aligned}\boldsymbol{\alpha}^{*\prime}\tilde{\mathbf{u}}_{mi} &= \boldsymbol{\alpha}'\mathbf{u}_{1i} + \boldsymbol{\alpha}'_d(\mathbf{u}_{mi} - \mathbf{u}_{1i}) + \alpha_{ns}(m-1) + \alpha_s t_{(m-1)i} \\ \boldsymbol{\beta}^{*\prime}\tilde{\mathbf{v}}_{mi} &= \boldsymbol{\beta}'\mathbf{v}_{1i} + \boldsymbol{\beta}'_d(\mathbf{v}_{mi} - \mathbf{v}_{1i}) + \beta_{ns}(m-1) + \beta_s t_{(m-1)i}.\end{aligned}$$

When covariates are recorded only at the beginning of the study, $\mathbf{v}_{mi} - \mathbf{v}_{1i} = \mathbf{0}$ and $\mathbf{u}_{mi} - \mathbf{u}_{1i} = \mathbf{0}$. Economou *et al.* list various further simplifications [ECO 15].

Model fitting and inference follow along the lines indicated above for the model of Pennell *et al.* As before, the contribution of a censored observation in the final stage is represented in the likelihood by the pdf of its unknown current level instead of the inverse Gaussian survival function. Prior distributions take a similar form to those in the earlier paper. A small difference arises in the MCMC procedure where an improved criterion is used at the accept-reject step.

The model is used by Economou *et al.* to analyze a dataset that records applications for treatment by 1497 drug users in Greece after 2001 [ECO 15]. The event of interest is "a new application for treatment". The hypothesized underlying latent stochastic processes can be interpreted as representing the psychological and physical status of the drug user; his or her new application for treatment occurs when the process has fallen to the threshold level at zero. For each individual, the time origin is at the first application after 2001. The first event time t_{1i} is the time of his or her next application, if any, and so on. The maximum number of events (equivalent to the maximum number of uncensored stages) was eight and the average number of stages per person was 1.43. Data collection ended in 2011 at which point all remaining times (final stages) were censored.

Five covariates were measured at the beginning of the study: place of residence (Athens/Piraeus, Thessaloniki or other areas); gender; age in years; the primary substance of abuse of the drug user (heroin or other); year of initial registration in the database, recoded as years since the establishment of the dataset by subtracting 2001. Since no information is available on which covariates affect the drift and which influence the starting point of the process, all the covariates were allowed to enter the predictors of both these characteristics of the process. Additionally, apart from the above five covariates which are time invariant, the number of previous stages and the total time since entry to the database were included. Consequently, the model contains 19 parameters in total (nine for the drift, nine for the initial stage and one for λ, the inverse of the variance of b_{mi}).

Figure 4.1 shows an example of the output for one of the parameters in the course of the MCMC procedure and a smooth kernel estimate of the density of the final posterior distribution. Table 4.1 presents the descriptive statistics of the posterior distribution of the parameters.

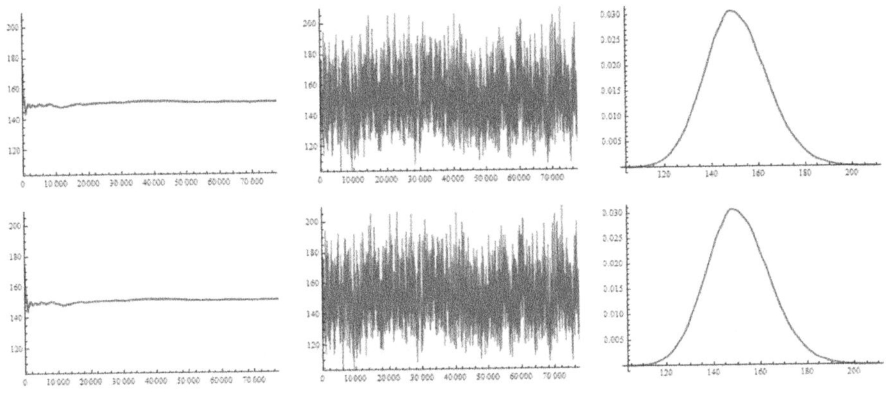

Figure 4.1. *The ergodic mean (left plot), the trace (middle plot) and a smooth kernel estimate of the density of the posterior distribution (right plot) of λ of the model fitted to the data on drug users [ECO 15]*

Residence in Greater Athens has a negative effect on the starting point, so the expected starting point for drug users living there seems to be lower than the corresponding value for the other two groups. The other dummy variable for place of residence, Thessaloniki, seems to have no significant effect on the starting point. Men tend to have a higher starting point than women, but this is barely significant. There is clear evidence that the older the age of a drug user, the lower his or her starting point. People who primarily use heroin have a much lower starting point compared to people who primarily use other drugs.

The year of initial registration in the database (variable Year of entry) has a positive coefficient for the starting point, meaning that later entries into the database have a higher starting point, that is, the new entrants in recent years tend to have higher initial psychological and physical status than entrants in the earlier years. The previous number of stages also has a positive effect on the starting point, meaning that the expected starting point of a drug user is higher than the expected starting point of his or her previous stage. One interpretation of this is that it shows a positive effect of treatment. The total time of the previous stages has a similar effect, although only marginally significant, since a longer period under observation suggests a higher expected starting point for the next stage.

	θ	Mean	Mode	Q_1	Median	Q_3	$P(\theta < 0 \mid \text{Data})$
starting point y_0	constant	21.711	21.948	19.925	21.660	23.384	<0.0001
	Greater Athens	–3.508	–3.853	–4.621	–3.520	–2.225	0.984
	Thessaloniki	–0.311	–0.619	–1.420	–0.303	0.947	0.566
	Male gender	1.086	1.114	0.627	1.097	1.555	0.059
	Age	–0.173	–0.170	–0.193	–0.172	–0.152	>0.9999
	Heroin	–5.338	–5.189	–6.271	–5.276	–4.359	0.999
	Year of entry	1.288	1.307	1.140	1.289	1.424	<0.0001
	No. of previous stages	1.670	1.708	1.246	1.669	2.074	0.001
	Total time	0.002	0.002	0.001	0.002	0.003	0.057
drift μ	constant	0.034	0.031	0.019	0.034	0.050	0.070
	Greater Athens	0.032	0.031	0.023	0.032	0.041	0.007
	Thessaloniki	–0.023	–0.022	–0.031	–0.023	–0.014	0.962
	Male gender	–0.003	–0.004	–0.009	–0.003	0.004	0.605
	Age	$9.1 \cdot 10^{-4}$	$9.0 \cdot 10^{-4}$	$6.1 \cdot 10^{-4}$	$9.0 \cdot 10^{-4}$	0.001	0.015
	Heroin	–0.010	–0.008	–0.018	–0.009	–0.002	0.791
	Year of entry	0.010	0.010	0.009	0.010	0.011	<0.0001
	No. of previous stages	0.058	0.057	0.053	0.058	0.063	<0.0001
	Total time	$9.7 \cdot 10^{-6}$	$9.4 \cdot 10^{-6}$	$2.9 \cdot 10^{-6}$	$9.6 \cdot 10^{-6}$	$1.64 \cdot 10^{-5}$	0.168
	λ	150.769	148.093	141.72	150.31	159.385	

Table 4.1. *Descriptive statistics for the posterior distributions of the parameters of the model fitted to the data on drug users (based on [ECO 15])*

Concerning the drift of the latent process, it appears that between-subject variability is explained adequately by the measured covariates and hence the estimated variance of the random effects is very small. Summarizing the results of Table 4.1, residence in Greater Athens, age, years since the start of the database and number of previous stages all have positive coefficients. Consequently, higher values of drifts - and hence a tendency to longer times until the next event - are associated with residence in Greater Athens, older age, later entry to the database and more previous stages.

The phenomenon of "opposite signs" of the estimated coefficients for the same covariate in the regressions for starting values and drift arises in respect of residence in Greater Athens and age. Older drug users and residents of Greater Athens tend to have lower starting values (implying shorter times to the next event) but less drift towards the boundary (implying longer times to the next event). This is not necessarily a contradiction. It could mean, for example, that older drug users are better able to cope with their circumstances than younger drug users who are at the same level of the process.

Further extension of this model is provided by Malefaki *et al.* [MAL 15]. Censoring is allowed to occur in intermediate stages as well as in the final stage. This is a realistic need if other events can intervene to prevent the occurrence of the event of interest. For example, in the application to drug users who seek treatment, imprisonment would be such an event.

4.6. Multiple outcomes

The basic concept of the FHT model is that the occurrence of the event of interest is triggered when the level of the underlying stochastic process, having started from a positive value, falls to zero. For the Wiener process with positive drift, we have seen that this may never happen, leading to the apparent presence of long-term survivors or a cured fraction. It seems reasonable to ask whether another, upper boundary could be defined in some contexts, representing "cure". Of course, this would have practical interest only if reaching this boundary corresponds to another observable event.

A context in which this structure arises is described by Horrocks and Thompson [HOR 04]. They model a patient's length of stay in hospital, with three possible outcomes: death in hospital, healthy discharge from hospital or transfer to another institution. Concentrating on the first two outcomes, let the underlying health level at time zero (entry to hospital) be $Y(0) = 0$ and define upper and lower thresholds at $u > 0$ and $-l$ ($l > 0$). Healthy discharge corresponds to reaching u before reaching $-l$, and death in hospital corresponds to reaching $-l$ before reaching u. The stochastic process with two absorbing boundaries is a standard problem and the necessary results can be found in, for example, Feller [FEL 86]. The joint distribution of T, the length of stay (until either death or discharge) and outcome D (0 for healthy discharge, 1 for death in hospital) is called, by Horrocks and Thompson, the First Passage Two Barrier distribution and its density is

$$f_{T,D}(t,j) = [f_u(t)]^{1-j} [f_{-l}(t)]^j ,$$

where $f_u(t)dt$ is the probability of hitting the upper barrier at u in the interval $[t, t+dt)$, without having reached $-l$ previously, and $f_{-l}(t)$ is defined similarly. Note that f_u and f_{-l} are subdensity functions that integrate to less than 1. Expressions for these subdensities as sums of infinite series involving normal densities are given by Horrocks and Thompson, as well as the expression for the survival function $S(t)$ which is required in the event of right censoring, when the patient is still alive in hospital at the end of the study. $S(t)$ is expressed as the sum of an infinite series involving normal integrals. Each of these functions depends on the boundary parameters u and l and also on the drift μ of the underlying Wiener process.

The log likelihood can be written as

$$\sum_{i:D_i=0} \ln [f_u(t_i)] + \sum_{i:D_i=1} \ln [f_{-l}(t_i)] + \sum_{i:D_i=2} \ln S(t_i),$$

where the outcome indicator D_i for the ith individual takes the value 0 for healthy discharge, 1 for death in hospital and 2 for right censoring (assumed uninformative as usual). Numerical optimization is feasible because the infinite series converge very rapidly.

The above methodology continues to apply when covariates are introduced, for example by $u_i = g_1(\boldsymbol{\beta}_u' \mathbf{x}_{ui})$ and $l_i = g_2(\boldsymbol{\beta}_l' \mathbf{x}_{li})$, where $\mathbf{x}_{ui}$ and $\mathbf{x}_{li}$ are the vectors of covariates affecting the upper and lower boundaries, respectively, $\boldsymbol{\beta}_u$ and $\boldsymbol{\beta}_l$ are the corresponding vectors of coefficients to be estimated, and g_1 and g_2 are suitable link functions.

In the illustrative example given by Horrocks and Thompson, the method is applied to the analysis of 100 elderly (aged at least 75 years) renal failure patients who either died in hospital ($n = 22$) or were discharged home ($n = 78$) with no censored observations. One covariate is considered, namely, a binary severity score x_i (value 1 for high severity, indicating major comorbidities and complications, and 0 otherwise). The upper and lower boundaries are modeled as linear functions of severity:

$$u_i = \beta_{u0} + \beta_{u1}x_i$$
$$l_i = \beta_{l0} + \beta_{l1}x_i,$$

where $x_i = 1$ if individual i has a high severity score and $x_i = 0$ otherwise.

The positive estimated value of $\beta_{u1} = 0.62$ indicates that a severe condition increases the distance to the upper barrier, so that discharge home is harder to achieve. On the other hand, $\beta_{l1} = -0.55$ indicates that the lower barrier (death in hospital) at $-l$ is closer and therefore easier to reach for patients with a severe condition. Methods suggested for assessing the goodness-of-fit of the model include the comparison of observed and expected cumulative incidence functions by severity group and outcome, and calculation of the Moran statistic by outcome (see Cheng and Stephens [CHE 89]).

In addition to heterogeneity modeled by dependence on measured covariates, individual heterogeneity can also be introduced by allowing the drift parameter μ to be drawn from a normal distribution. Furthermore, the third possible outcome, transfer to another institution, can be included within this framework by assuming a third barrier at w, taking $0 < w < u$ if the context requires that transfers should be modeled as occurring when health has improved or $0 > -w > -l$ if transfers occur when health deteriorates. When the patient's health status first reaches w, without having reached either u or $-l$, a transfer takes place with probability p. Otherwise, the patient remains in the hospital until the path reaches u or $-l$ (or is censored).

As discussed by Horrocks and Thompson, the above analysis of length of stay is a competing risks analysis, but quite different in concept from the classical approach that assumes latent failure times $T_1, \ldots, T_c$ for each of c causes of failure, with observed failure time $T = \min T_i$ (see, for example, Crowder [CRO 01]; this approach is used in the competing risks model that appears in the next section). They describe it as being closer to the approach of Kalbfleisch and Prentice [KAL 02] who model separate subhazard functions for the various causes. However, in the present model, the presence of all of its parameters in every one of these functions ensures dependence between the outcomes. Thus, once again, we see that an FHT model based on an underlying stochastic process succeeds in providing a natural representation of a feature of the real-world situation that is difficult to capture in other lifetime data models.

A similar model for competing risks is proposed by Lindqvist and Skogsrud, who envisage a situation in which the working lifetime of a component or machine is terminated either by failure or by preventive maintenance [LIN 08]. The necessity for the latter arises when the underlying Wiener degradation process (starting from zero) first reaches a boundary $s > 0$, which is below the critical threshold c at which failure occurs. This is similar to the extension of the model of Horrocks and Thompson that adds a boundary w at which a patient may be transferred to another institution, before the boundary $-l$ denoting death is reached. Just as Horrocks and Thompson propose that the transfer takes place with a probability $p < 1$, so Lindqvist and Skogsrud suppose that the signal that the process has reached s is not necessarily detected. If it is not detected, no maintenance takes place and the process continues until c is reached, whereupon the machine fails. Note that the two competing risks - maintenance and failure - are dependent, because of their common dependence on the underlying degradation process, which is appropriate for this reliability problem. The underlying process here is failure-defining in Cox's classification given in section 2.11 [COX 99].

4.7. Extensions of the basic FHT model in a study of low birth weights: a mixture model and a competing risks model

Two extensions of the basic FHT regression model based on an underlying Wiener process are presented by Whitmore and Su in a study of low birth weights in the USA [WHI 07]. The issue of low birth weight (conventionally meaning below 2,500 g) is important because these newborns have increased risks of infant mortality and other adverse outcomes. Babies born preterm (conventionally, under 37 weeks from conception) are similarly at risk. Earlier analyses of these phenomena by logistic regression (with dependent variable the yes/no classification of low birth weight or short gestational age) or linear regression (treating birth weight as a continuous response) are replaced here by models based on conceptualizing the process of fetal development. Analysis is carried out using a 4% sample of the live births in the USA in 2002, which still amounts to over 160,000 births. Many covariates are included in the database that was available for this research, but only four are used here for the purpose of illustrating the methodology. These are the following: mother's age; mother's race; number of previous live births; method of delivery.

Let $\mathbf{D}(t)$ represent a multidimensional underlying fetal development process, including observable components (such as weight) and latent or unknown components (such as the development of the brain). Assume that birth occurs when $\mathbf{D}$ first reaches a birthing boundary $\mathcal{B}$ in the multidimensional space. Thus, the gestational age is

$$T = \min \{t : \mathbf{D}(t) \in \mathcal{B}\}$$

and the birth weight is $W = D_1(T)$ if birth weight is the first component of $\mathbf{D}$. Two models are studied.

4.7.1. *Mixture model*

Let the development process be two-dimensional:

$$\{\mathbf{D}(t)\} = \{D_1(t), D_2(t)\} \text{ with } \mathbf{D}(0) = \mathbf{0},$$

where $D_1(t)$ is the fetal weight at time t and $D_2(t)$ represents some other latent development measure. Define a unidimensional latent fetal development process $C(w)$ as a function of weight $w = D_1(t)$ at time t: that is,

$$\{C(w)\} = \{C(D_1(t))\} = \{D_2(t)\}.$$

Further assume that $C(w)$ is a Wiener process with mean parameter μ and unit variance, and that the birthing boundary $\mathcal{B}$ represents a fixed level $b > 0$ of development. Thus, birth occurs when $C(W) = b$ for the first time. So far, this is the framework of a standard application of the methodology that was introduced in Chapter 2 (with the minor difference that it starts from zero and increases towards a boundary rather than decreasing from its initial value to a boundary located at zero). Now, however, suppose that in a proportion p_0 of cases, the birthing process diverges from the natural path of healthy development. In these cases, development still takes the form of a Wiener process but with different parameters and a different boundary. Then, the overall pdf of birth weight takes the form of a mixture of two inverse Gaussian distributions:

$$f(w) = p_1 f_1(w) + p_0 f_0(w),$$

where the two pdf's f_1 and f_0 denote the natural and divergent paths, respectively, and $p_1 = 1 - p_0$. Whitmore and Su [WHI 07] cite earlier work that similarly uses a two-component mixture distribution, but mixes two normal distributions and lacks the regression structure that is imposed on the parameters of the pdf's f_1 and f_0 in the present study. This structure assumes the usual identity link functions for the mean parameters μ_i and logarithmic for the boundaries b_i $(i = 0, 1)$ and the logit function $\ln [p_1/(1 - p_1)]$ for the mixing proportion p_1.

4.7.2. *Competing risks model*

In an alternative conceptualization of the birthing process, fetal development is taken to be a three-dimensional process, with fetal weight $w = D_1(t)$ as its first component. Consider the two-dimensional subordinated process $\{C_1(w), C_0(w)\}$ where normal development is represented by movements along the dimension $C_1(t)$ but abnormal development is represented by the other dimension $C_0(t)$. These two components are taken to be uncorrelated Wiener processes, with means μ_i and fixed boundaries b_i as before. Then, the birth weight is the weight realized at the smaller of the two first hitting times on these boundaries:

$$W = \min\{W_1, W_0\}, \text{ where } W_N = \min\{u : C_N(u) \geq b_N\}, \text{ for } N = 1, 0$$

and the pdf of birth weight is

$$f(w) = f_1(w)\left[1 - F_0(w)\right] + f_0(w)\left[1 - F_1(w)\right].$$

The parameters μ_i and b_i depend on covariates in the same way as in the mixture model. The competing risks model, lacking the two parameters p_i, is more parsimonious than the mixture model.

Note that the second model, competing risks, represents the outcome (belonging to the low or normal birth weight group) as depending on the course of the birthing process. In contrast, the first model, the mixture model, implied that it was determined from the moment of conception whether the birthing process would take the normal or abnormal path. This significant issue in the conceptualization of models was mentioned, in a slightly different but related context, in our discussions of models for cured fractions or long-term survivors earlier in this book.

4.7.3. *Comparative results*

Although the analysis presented by Whitmore and Su was intended only for illustrative purposes, some features of the results are worthy of comment. The mixture model gives an estimate of p_0, averaged over covariate values, of 6.5%. (In comparison, 7.8% of actual birth weights fall below the conventional 2500g cut-off.) In the competing risks model, the sign of the estimated mean parameter μ_0 in about 85% of cases indicates drift away from the abnormal birthing barrier b_0, and hence a non-zero probability that this barrier would never be reached. To a large extent, it is this feature that determines the high percentage of normal births.

The parameter estimates obtained for the much larger component of normal birth weights give the inverse Gaussian distribution a shape very close to the normal distribution. The corresponding part of the distribution from the competing risks model is very similar, but more noticeable differences arise in the part corresponding to low birth weights. Both models seem to predict more low birth weights than actually appear in the data. This discrepancy may arise because only live births were considered. A further extension to the competing risks model would include still births and spontaneous abortions as further alternative outcomes of the course of pregnancy.

A paper containing further development of the topic of birth weights will be mentioned in section 6.8.

Another model with some features in common with the above is the competing risks model presented by Xu *et al.* [XU 11]. One component of the two-component mixture model, representing lifetimes terminated by the cause of interest, is an inverse Gaussian distribution. The other, representing "cured" individuals who do not die from the cause of interest, is a Gompertz distribution. The purpose of this structure is to allow for deaths from other causes among the non-susceptible individuals who would otherwise appear as long-term survivors. The Gompertz distribution is used because it is assumed to represent risk in the general population (see section 1.4). The model is illustrated on Boag's data on breast cancer (see section 1.17) and data on melanoma that are referred to elsewhere in this book (see sections 2.7, 4.4 and 4.10).

4.8. Semi-parametric modeling of covariate effects

The extension to the basic FHT regression model in a study by Yu *et al.* [YU 09] is motivated by an analysis of data on sexually transmitted infections (STI) in adolescent women. The authors regard the Cox PH regression models that are traditionally used for time-to-event analysis in biomedical applications as having clear limitations in this setting. For example, adolescent women are likely to change partners and are thereby exposed to different levels of risk, so the assumption that hazard ratios are constant seems implausible. The authors instead adopt FHT regression modeling for this problem and furthermore aim to increase flexibility by extending the model in order to incorporate semi-parametric components with the purpose of capturing possible nonlinear covariate effects.

The structure and parameterization adopted for the FHT model based on an underlying Wiener process is as follows. First, the parameterization adopted for the inverse Gaussian distribution is

$$f(y|\mu,\nu) = \left(2\pi y^3 \nu\right)^{-1/2} \exp\left(-\frac{(1-\mu y)^2}{2\nu y}\right)^2, \ y > 0$$

and the regression model $\mu = \theta(z) + \gamma'\mathbf{x}$ is adopted for μ the reciprocal of the mean survival time. Here, the unspecified smooth function $\theta(z)$ of the covariate z is added to the linear effects of the covariate vector $\mathbf{x}$.

For an observed event, the contribution to the log likelihood is

$$\begin{aligned}\ell_i(\theta,\gamma,\nu) &= \ln\left\{f\left(y_i|\theta,\gamma,\nu\right)\right\} \\ &= -\frac{1}{2}\ln(2\pi\nu) - \frac{3}{2}\ln(y_i) \\ &\quad - \frac{1 - 2\left\{\theta(z_i) + \gamma'\mathbf{x}_i\right\} y_i + \left\{\theta(z_i) + \gamma'\mathbf{x}_i\right\}^2 y_i^2}{2\nu y_i}\end{aligned}$$

and, for a censored observation, the contribution is

$$\ell_i^c(\theta,\gamma,\nu) = \ln\left\{P(Y_i \geq y_i)\right\} = \ln\left\{\int_{y_i}^{\infty} f(s|\theta,\gamma,\nu)ds\right\}.$$

Regression splines are used to model the nonlinear covariate effect, specifically cubic B-splines. Thus,

$$\theta(z) = \sum_{j=1}^{m} \beta_j B_j(z).$$

Finally, the log likelihood becomes

$$\ln\{f(y_i|\boldsymbol{\beta},\gamma,\nu)\} = -\frac{1}{2}\ln(2\pi\nu) - \frac{3}{2}\ln(y_i) - \frac{1 - 2\sum_{j=1}^{m}\mu_{ij}(\boldsymbol{\beta},\gamma)y_i + \sum_{j=1}^{m}\mu_{ij}(\boldsymbol{\beta},\gamma)^2 y_i^2}{2\nu y_i},$$

where $\mu_{ij}(\boldsymbol{\beta},\gamma) = \beta_j B_j(z_i) + \gamma'\mathbf{x}_i/m$. This is regarded as the log likelihood of data with covariates $\{B_1(z_i), B_2(z_i), \ldots, B_m(z_i), \mathbf{x}_i'\}'$ and corresponding coefficients $(\beta_1, \beta_2, \ldots, \beta_m, \gamma')'$. A pair of estimating equations are derived which can be solved by a simple iterative procedure.

An alternative method uses penalized splines with the intention of obtaining a smoother estimate of the non-parametric function. Adding a penalty term gives the penalized log likelihood

$$\sum\{\delta_i\ell_i + (1-\delta_i)\ell_i^c\} - \frac{\lambda}{2}\boldsymbol{\beta}'\mathbf{P}\boldsymbol{\beta},$$

where the penalty matrix $\mathbf{P}$ is known. Estimation follows a course closely similar to the unpenalized case. The choice of the number of knots and the smoothing parameter for the spline estimates are discussed and a simulation study is carried out to demonstrate the reliable performance of the proposed method. The results suggest an advantage to using penalized rather than unpenalized splines, in terms of smaller bias and standard errors.

The application is to the occurrence of $C.\ trachomatis$ infection in 387 women (aged 11–17 years) enrolled in three adolescent medicine clinics. Three covariates are considered: age at first intercourse, number of episodes of unprotected sexual intercourse in the last three months, and lifetime number of sexual partners (at the time of enrolment). In Cox regression, the

PH assumption is rejected globally and also specifically for age at first intercourse. In the new approach outlined above, a non-parametric effect for the lifetime number of sexual partners is introduced. It turns out that infection risk is low for one, two or three partners, markedly higher for four to seven partners, and lower again for more than seven partners. This shape - previously unreported in the literature - can be interpreted as describing a pattern in which the adolescent females with few partners are usually younger and seeing correspondingly young male partners, who are an unlikely source of infection. At the other end of the scale, adolescent women with many partners may be more mature and aware of STI issues, and consequently more cautious in partner selection.

The details of the method are presented for the case in which only one covariate's effect is to be modeled non-parametrically, but the approach is not limited to that situation.

4.9. Semi-parametric model for data with a cured fraction

Another model that introduces a semi-parametric structure into the Wiener-based lifetime model is proposed by Choi *et al.* [CHO 14], with an emphasis on the inverse Gaussian model's ability to represent a cured fraction. An underlying Wiener process $W(t)$ is assumed to have the drift $-\mu A(t)$ and the variance parameter $A(t)$, with $W(0) = 0$ and $A(t)$ a non-decreasing function which will not be specified parametrically. Dependence on covariates $\mathbf{x}$ is introduced by replacing $A(t)$ by $A(t)e^{\boldsymbol{\beta}'\mathbf{x}}$. Furthermore, the threshold α at which the event occurs is also allowed to depend on covariates, $\alpha = e^{\gamma'\mathbf{x}}$. The survival model is

$$S_i\left(t|\mathbf{x}_i\right) = P\left(T_i \geq t|\mathbf{x}_i\right) = G_\alpha\left[\int_0^t Y_i(s)\exp\left\{\boldsymbol{\beta}'\mathbf{x}_i(s)\right\}dA(s)\right],$$

where $G_\alpha(t)$ denotes the inverse Gaussian distribution with parameters α and $\mu = 1$ (the latter being required for model identification in this non-parametric context). This model arises as the first hitting time on the boundary at α of the non-homogeneous process $W(t) = W_0(\Lambda_{\mathbf{x}}(t))$, where $W_0(t)$ denotes a homogeneous Wiener process with first hitting time T_0 on this boundary and $\Lambda_{\mathbf{x}}(t) = \int_0^t Y(s)\exp\left\{\boldsymbol{\beta}'\mathbf{x}(s)\right\}dA(s)$. Thus, the model incorporates a flexible time scale for the deterioration of the underlying state of health.

The model is fitted by a non-parametric maximum likelihood algorithm and simulation studies demonstrate the performance of the model. The illustrative application is to a set of data on 674 patients with soft tissue sarcoma (STS) who received either local therapy only or local therapy plus chemotherapy. Kaplan-Meier survival curves suggest a considerable proportion of long-term survivors. Furthermore, the Kaplan-Meier curves for the two treatment groups (with deaths from causes other than STS treated as censored) crossed over at about 20 months. As this feature is incompatible with a PH model, the original analysis of these data was based on two Cox analyses, one for the first year and one thereafter. This approach is questionable for many reasons. In the modeling within the framework of the FHT model, the coefficients $\boldsymbol{\beta}$ referred to the treatment plus four other covariates, while γ referred only to the treatment effect. The fit of the model was ultimately rather similar to the fit of two models that were not based on an underlying process, namely the mixture model for a cured fraction and the promotion time cure rate model (see Tsodikov *et al.* [TSO 03]), both of which incorporated a Cox model for survival times.

4.10. Semi-parametric time-varying coefficients

It is sometimes necessary to enhance a statistical model by allowing the regression coefficients to vary over time. For example, this might make it possible to model the effect of a baseline covariate that has a strong early effect that later diminishes; see Peng and Huang [PEN 07], for example. An analysis by Li and Lee [LI 11] investigates such a situation. They assume the FHT model based on an underlying Wiener process, with the resulting inverse Gaussian distribution of survival times parameterized as

$$f(y|\mu,\nu) = \left(2\pi y^3 \nu\right)^{-1/2} \exp\left\{-\left(1-\mu y\right)^2/(2\nu y)\right\},\ y>0,$$

as in section 4.8.

The following regression structure is adopted for the parameter μ:

$$\mu_i = \boldsymbol{\alpha}(U_i)'\mathbf{x}_i + \boldsymbol{\beta}'\mathbf{z}_i,$$

where both $\mathbf{x}_i$ and $\mathbf{z}_i$ represent covariates. The regression coefficients $\boldsymbol{\beta}$ of $\mathbf{z}_i$ are constant but those of $\mathbf{x}_i$ are given by the functions $\boldsymbol{\alpha}$ that vary according to a further covariate, U. These functions are not specified. Without the term in $\mathbf{x}_i$, this is simply the regression model introduced by Whitmore [WHI 83] and if $\mathbf{x}_i$ involves only an intercept, it is the model of Yu *et al.* [YU 09] presented in section 4.8. Thus, the new model generalizes previous approaches. The variable U is likely to be related to time - in the application presented in the article, it is the subject's age - but does not have to be.

The sketch of the estimation procedure given by Li and Lee [LI 11], before they go into details, is as follows. A local polynomial approach is adopted, so that for U_i close to a point u,

$$\begin{aligned}\boldsymbol{\alpha}(U_i) &\approx \boldsymbol{\alpha}(u) + (U_i - u)\boldsymbol{\alpha}^{(1)}(u) + \ldots + (U_i - u)^k \boldsymbol{\alpha}^{(k)}(u)/k! \\ &= \{\mathbf{u}_i(u) \otimes \mathbf{I}_p\}' \boldsymbol{\theta}(u),\end{aligned}$$

where $\mathbf{u}_i(u) = \left(1, (U_i - u), \ldots, (U_i - u)^k\right)'$, $\boldsymbol{\theta}(u) = (\boldsymbol{\alpha}(u)', \ldots, \boldsymbol{\alpha}^{(k)}(u)'/k!)'$. Consequently, the log likelihood of the observed data takes the form

$$\ell_u = \sum_{i=1}^{n} \{\delta_i \ell_i + (1 - \delta_i)\ell_i^c\} K_h(U_i - u),$$

where $K_h(.) = K(./h)/h$ for kernel function K and h is a bandwidth. Here, ℓ_i is an observed survival time's contribution to the log likelihood and ℓ_i^c is a right-censored time's contribution, with δ_i the usual censoring indicator. This equation can be maximized to obtain $\hat{\boldsymbol{\theta}}(u)$ and its first p entries give the local maximum likelihood estimates $\hat{\boldsymbol{\alpha}}(u)$ of $\boldsymbol{\alpha}(u)$. Repeating this computation for different values of u over the range of U yields the complete profile of values of $\boldsymbol{\alpha}(u)$. The full estimation procedure is supported by a theoretical analysis and a simulation study.

The application presented is to the set of data on malignant melanoma, which form a suitable example for a new method because it has been noted in the literature that the PH assumption may not hold. For example, Keiding *et al.* [KEI 97] found it necessary to introduce a frailty term to represent additional unobserved heterogeneity (see section 1.18). Regression coefficients varying with age are introduced for three covariates (and the intercept also varies), with a variety of patterns emerging, all clearly not constant. A generalized likelihood ratio test is used to test one-by-one the hypotheses

$$H_0 : \alpha_k(.) = \alpha_{k0} \ \ vs \ \ H_1 : \alpha_k(.) \neq \alpha_{k0}$$

of constant coefficients, using a bootstrap procedure to obtain the null distribution of the test statistics. All the null hypotheses are rejected with $P < 0.01$.

4.11. Bivariate Wiener processes for markers and outcome

Making the extension to a bivariate Wiener process provides a way of incorporating a marker process (see section 2.11) into the model. This marker process is an observable process indicating degradation (of the individual's health, for example) which is correlated with the latent process (the individual's health status). However, failure (death, relapse) is determined by the latter, not the former. In this respect, the structure that is envisaged is different from work in reliability in which failure is usually determined by the observable marker process, for example, the measurable growth of a crack.

Whitmore *et al.* set up the following model for this situation [WHI 98]. One component of the bivariate Wiener process is a degradation process $Z(t)$, with $Z(0) = 0$; the unit fails when Z first reaches a boundary at $a > 0$. The other component is a marker process $Y(t)$, with $Y(0) = 0$, which is correlated with Z and tracks its progress. The vector $(Z(t), Y(t))$ has a bivariate normal distribution with mean $t\,\boldsymbol{\mu} = (t\,\mu_z, t\,\mu_y)$ and positive definite covariance matrix $t\,\boldsymbol{\Sigma}$. It is assumed that the value of the marker process is recorded only at the moment of failure or censoring. Hence, two probabilities are required in order to write down the likelihood of a set of data incorporating possible right censoring:

– for a surviving item censored at time t, $P\,(Y(t) = y, S > t)$,

– for an item failing at time s, $P\,(Y(s) = y, S = s)$,

where the random variable S is the failure time and y is the measurement of the marker process at failure or censoring. Because of the bivariate normality of the Wiener process, and the inverse Gaussian distribution of S, these probabilities can be written as expressions that involve nothing more elaborate than the normal distribution function. In the event of a complete sample (no censoring), estimates of the parameters $\boldsymbol{\mu}$ and $\boldsymbol{\Sigma}$ can be written in explicit forms, and standard distributional properties of normal and inverse Gaussian samples lead to straightforward construction of confidence regions for parameters.

The formulation of the latent process by Whitmore *et al.* was suitable for degradation increasing from zero to a fixed threshold, as is usually appropriate in reliability contexts. The further development of the model by Lee *et al.* [LEE 00] used a framework more suitable for health-related processes, with $Z(0) > 0$ and failure occurring when $Z(t)$ first falls to zero. The marker process $Y(t)$ is defined in terms of changes from the baseline of a process $Y_w(t)$, thus $Y(t) = Y_w(t) - Y_w(0)$. Furthermore, the model is extended to incorporate fixed covariates $\mathbf{x}$ measured at baseline in the usual way, that is, by allowing every one of its parameters to depend on a linear combination $\beta'\mathbf{x}$ through suitable link functions. One of the parameters is the correlation ρ between the processes, for which the link function

$$\left(e^{\beta'\mathbf{x}} - 1\right) / \left(e^{\beta'\mathbf{x}} + 1\right)$$

is suggested in order to impose the necessary restriction to the range $(-1, +1)$. The application of the model is illustrated using data from a large ($n = 787$ patients) clinical trial of antiretroviral drugs for the treatment of AIDS. The authors argue that the development of antiviral resistance will reduce the effect of drugs over time, so that a proportional hazards model must be ruled out. The marker process is the change in the CD4 cell count (on a log scale) from its baseline value. The baseline value is included among the covariates. Among the results, the estimated correlation between marker process and health status was found to be 0.242 - significantly different from zero, but not large enough to suggest that the marker has great value in tracking the patient's progress.

Model fit is assessed by visual comparison of fitted survival curves to Kaplan-Meier estimates. Techniques for checking the assumptions of this model are discussed further by Lee and Whitmore in a separate paper [LEE 02a]. Their suggestions include:

– checking the Wiener property of the latent disease process by means of an inverse Gaussian probability plot or comparing fitted survival curves to the corresponding non-parametric Kaplan-Meier estimate, as seen in Figure 3 of Lee *et al.* [LEE 00].

– checking the Wiener property of the observable marker process by means of a normal probability plot in order to confirm that increments are normally distributed, as seen in Figures 1 and 2 of Lee *et al.* [LEE 00].

– checking the assumption of independent marker increments by checking the correlation between successive increments recorded for the same subject.

Denote one such increment by

$$\Delta y_{ij} = Y_{wi}\left(t_{ij}\right) - Y_{wi}(t_{i,j-1}),$$

where the process for subject i was observed successively at times $t_{i,j-1}$ and t_{ij}. Then, the correlation between pairs $(\Delta y_{ij}, \Delta y_{i,j-1})$ should be zero if increments are independent. In practice, the marker process is likely to be measured with error, which means that the presence of $Y_{wi}(t_{i,j-1})$ in both Δy_{ij} and $\Delta y_{i,j-1}$ induces a small negative correlation between them.

– examining goodness-of-fit for survival times; this suggestion was already outlined in section 3.8.

Potential remedies in case deficiencies identified in the fit of the model include: transformations, for example transformation of the calendar time scale (see, for example, [WHI 97]) and taking competing risks into consideration.

Note that both the model introduced by Whitmore *et al.* and this extension by Lee *et al.* assume only one measurement of the marker process. This is made at the end of the individual's participation in the study, that is, at the time of the event or of censoring. Lee *et al.* propose a further extension to allow repeated measurements of the marker process, by treating the increments over the time intervals in between measurements as independent. A more complete modeling approach to longitudinal data is given by Hashemi *et al.*, with the limitation of assuming a linear trend in time of the latent process [HAS 03]. Another more elaborate approach to longitudinal data can be found in section 6.13.

5

Relationship of First Hitting Time Models to Proportional Hazards and Accelerated Failure Time Models

5.1. Introduction

Among the models that belong to the class of first hitting time regression models, the emphasis in this book falls on the inverse Gaussian model based on an underlying latent Wiener process. As with much of the work on lifetime data, its applications are to be found chiefly within the biomedical sciences. The predominant model in that area is, of course, Cox's semi-parametric proportional hazards model. Therefore, it is natural and indeed necessary to compare and contrast first hitting time (FHT) and proportional hazard (PH) models. The investigation should also be extended to include the other major category of regression models, i.e. accelerated failure time (AFT) models.

We first review some direct comparisons between FHT and PH models applied to the same sets of data. Subsequently, we examine published studies of the theoretical connections between the various classes of models.

5.2. FHT and PH models: direct comparisons by case studies

5.2.1. *Case study 1: mortality after cardiac surgery*

After a comparative discussion of the various properties of PH and FHT models, Stogiannis *et al.* [STO 11] proceeded to carry out direct comparisons by fitting Cox semi-parametric PH regressions and inverse Gaussian FHT regressions to the same data. This is done both for simulated data, for which the correct model is known, and for a large set of real data with many available covariates. The aim of analyzing the simulated data is to illustrate what might happen when the inverse Gaussian FHT model is fitted to data that satisfy a PH model and vice versa. The results were described briefly in section 3.9.3.

The real data in the case study that was examined by Stogiannis *et al.* was earlier analyzed using Cox's model by DeRose *et al.* [DER 05]. These data refer to the survival of 544 patients who had undergone coronary artery by-pass grafting (CABG). A total of 11 covariates were retained for the multivariable Cox model following the commonly adopted strategy of testing a large number of covariates in univariate Cox models and discarding those with no apparent association with survival. Subsequently, of these 11 covariates, only five were kept in the final Cox model after a backward elimination procedure. However, in the comparative analysis using inverse Gaussian FHT regression, all the 11 covariates were allowed to enter into the usual linear predictors for μ and $\ln y_0$. Four were statistically significant (at the 5% level) for the former and nine for the latter, including three significant for both (Table 5.1). One (past CHF) was significant for neither, and this was one of the five that had been selected by the Cox PH analysis. The variety of these results is even greater when the signs of the coefficients of the variables that are common to both linear predictors are taken into consideration. The age variable had negative coefficients in both, indicating that the older patient fares worse both in terms of faster drift towards the boundary at zero as well as a lower starting level. However, the covariates current CHF and COPD had opposite signs in the two predictors; this phenomenon was discussed in section 3.9.

In this case study, as is true for all analyses of real data, the correct model is not known. Diagnostics would help in order to avoid selection of an obviously wrong model. One major difference between the two models is the predicted presence of long-term survivors. In the inverse Gaussian FHT model, the fitted value of μ was positive for almost all the patients (536/544), who therefore have a non-zero probability of never reaching the boundary. Consequently, a high proportion of long-term survivors is expected. The overall Kaplan-Meier survival curve (Figure 5.1) suggests that this proportion could be approximately 40%, although the visual evidence is not absolutely convincing.

	Cox regression		FHT regression			
			$\ln y_0$		μ	
	Coeff	se	Coeff	se	Coeff	se
Age	0.049	(0.008)	–0.029	(0.004)	–0.003	(0.001)
Emergency surgery	0.580	(0.178)	–0.569	(0.126)		
Past CHF	0.430	(0.158)				
Peripheral vascular disease	0.507	(0.153)	–0.376	(0.094)		
COPD	0.366	(0.158)	0.952	(0.101)	–0.125	(0.022)
Body mass index			0.019	(0.010)		
Urgent surgery			0.773	(0.123)		
Current CHF			–1.018	(0.087)	0.083	(0.022)
Calcified aorta			–0.508	(0.108)		
Immune deficiency			1.334	(0.417)		
Off-pump surgery					0.089	(0.036)
Constant			1.739	(0.362)	0.351	(0.071)

Table 5.1. *Parameter estimates (where significant at the 5% level in Wald tests) from Cox regression and inverse Gaussian FHT regression [STO 11]. (COPD = chronic obstructive pulmonary disease; CHF = congestive heart failure)*

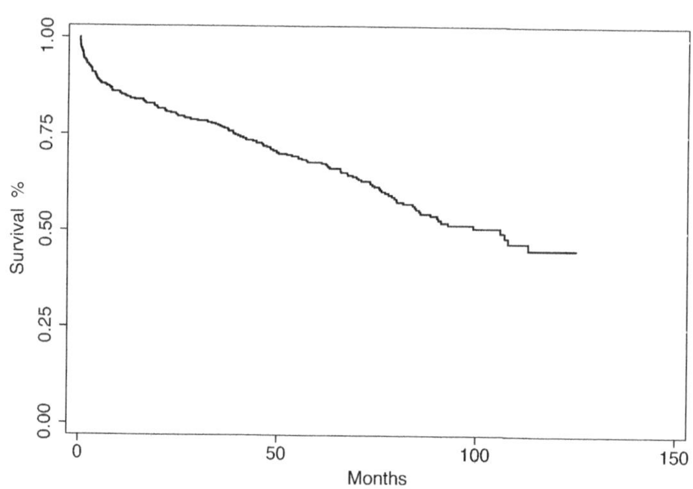

Figure 5.1. *Kaplan-Meier estimate of survival time in months after surgery of 544 CABG patients [DER 05, STO 11]*

5.2.2. *Case study 2: lung cancer in a cohort of nurses*

Another case study for comparing FHT and Cox PH regression models, published by Lee *et al.* [LEE 10c], is based on the data from a large cohort study, the Nurses' Health Study, in the USA. Observations for 115,768 women in the period 1986–2000 were included in their analysis, with the endpoint being the diagnosis of lung cancer. Only a restricted set of covariates were used for this comparison. Age at entry to the study (i.e. at time zero) and cumulative smoking at entry (in pack-years) were included in the linear predictors for both μ and $\ln y_0$. Additionally, the square of the cumulative smoking variable and a variable representing the average amount of smoking since entry to the study were included in the model for μ only (note that the latter variable logically can affect only μ, not y_0). Despite the enormous sample size, the P-value for age in the regression for $\ln y_0$ was only 0.066, compared to $P < 0.001$ for all the other parameters in the model. The same four covariates were then entered into a Cox PH regression model fitted to the same data. All were significant at $P < 0.001$, with signs of estimated coefficients that agreed with the results of the inverse Gaussian FHT regression. The application of the diagnostic test described in section 3.8 suggested that both models provided equally good fit to the data. A global test did not reject the hypothesis of proportional hazards (this might be thought a somewhat surprising result in such a large sample, perhaps reflecting on the power of the test).

In summary, and unlike the comparison that was discussed in the previous section, this case study did not find very large differences between the results of the two models. The authors commented that the duration of the study may have played a role in this. Sixteen years is not a long time for the appearance of lung cancer cases in a cohort and, in fact, only 1206 cases were recorded, that is, only 1% of the participants. In contrast, in the first case study described above, there were 192/544 events (35%).

5.3. FHT and PH models: theoretical connections

The usual presentation of the FHT approach to lifetime data emphasizes its important conceptual differences from the PH approach. In this light, it may sound surprising that Lee and Whitmore are able to show that, in fact, PH can be thought of as a special case of FHT for most purposes [LEE 10a]. The following paragraphs present their analysis of the connection between the two methodologies.

First, it should be observed that an FHT regression model comprises three building blocks:

– the underlying stochastic process that describes the development of the unit's state of health (or whatever concept is relevant to the particular application);

– the threshold or boundary: the process's first arrival at this critical level triggers the event of interest;

– the time scale.

Of these three blocks, the first formed part of the basic presentation of the FHT models in Chapter 2.

Each one of the three building blocks may depend on parameters, which in turn may depend on vectors of covariates x (not necessarily the same covariates for each parameter). Whitmore and Lee proceed to demonstrate that a PH model can be constructed by varying the time scale of an FHT regression model or by varying the boundary. It is left as an open question whether a potential third method, namely, varying the stochastic process for given boundary and time scale, can produce a non-trivial PH model.

5.3.1. *Varying the time scale*

As remarked in Chapter 1, the relevant scale of a "lifetime" may not be calendar or clock time t, but an operating time or running time or other appropriate measure. In general, this process time is $r(t|\mathrm{x})$, a non-decreasing function of t satisfying $r(0|\mathrm{x}) = 0$. However, the development in the article actually assumes that $r(t|\mathrm{x})$ is an increasing function of t. One example is the measure of disease progression in Lee *et al.* [LEE 04]: the disease (lung cancer) advances over time, but at a quicker or slower rate depending on how much a worker who switches between types of work within the railroad industry is exposed to health stressors in his or her current working environment. That example is discussed further in section 6.2.

Let $S_R(r)$ be the survival function defined in terms of process time. Then, the survival function in calendar time is

$$S(t|\mathrm{x}) = S_R(r(t|\mathrm{x})) \qquad [5.1]$$

and if the PH property holds in calendar time, then by definition

$$S(t|\mathbf{x}) = \exp\left[-\exp(\boldsymbol{\beta}'\mathbf{x})\int_0^t h_0(u)du\right] = \exp\left[-\exp(\boldsymbol{\beta}'\mathbf{x})H_0(t)\right]$$

for a baseline hazard function $h_0(t)$ and corresponding cumulative hazard $H_0(t)$.

Equating to S_R and solving for $r(t|\mathbf{x})$ gives

$$r(t|\mathbf{x}) = S_R^{-1}\left\{\exp\left[-\exp(\boldsymbol{\beta}'\mathbf{x})H_0(t)\right]\right\}. \qquad [5.2]$$

This construction establishes that for any PH model with cumulative baseline hazard H_0, there is a corresponding function in process time that generates it.

The following two examples are provided by Lee and Whitmore [LEE 10a].

Poisson process. The first example is motivated by supposing that recurrent infections occur during the course of a disease. These could arise from random exposures to infectious agents, so a Poisson process model is a possibility (see also the example on recurrent exacerbations of disease in chronic obstructive pulmonary disease, section 6.7). In process time, the survival function for a Poisson process with unit rate is $S_R(r) = \exp(-r)$. Therefore, from equation [5.2],

$$r(t|\mathbf{x}) = \exp(\boldsymbol{\beta}'\mathbf{x})H_0(t).$$

This implies that the PH family corresponds to a non-homogeneous Poisson process with the above rate function $r(t|\mathbf{x})$. The covariate vector x has the effect of accelerating or decelerating the rate at which time runs.

Brownian motion. Motivation for the second example comes from considering that the patient receives an intervention when an established infection reaches a threshold. If the starting level of the Brownian motion is $y_0 > 0$ and the boundary is at zero, then from previous results the survival function is

$$S_R(r) = 1 - 2\Phi\left(-y_0/\sqrt{r}\right),$$

where Φ is the standard normal distribution function, from which

$$S(t|\mathbf{x}) = 1 - 2\Phi\left(-y_0/\sqrt{r(t|\mathbf{x})}\right), \qquad [5.3]$$

and hence

$$r(t|\mathbf{x}) = \left(\frac{y_0}{\Phi^{-1}\left\{\frac{1}{2}\left(1 - \exp\left[-e^{\boldsymbol{\beta}'\mathbf{x}}H_0(t)\right]\right)\right\}}\right)^2.$$

5.3.2. *Varying the boundary*

Let the boundary be $b(t|\mathbf{x})$, depending on time as well as on the covariates. Let $Pr[b(t|\mathbf{x})]$ denote the probability that a sample path starting from the origin lies above this boundary at time t. $F(s|\mathbf{x})$ denotes the distribution function of the first hitting time of the sample path on the boundary. The key equation is

$$Pr\left[b(t|\mathbf{x})\right] = \int_0^t Pr\left[b(t|\mathbf{x}) - b(s|\mathbf{x})\right] dF(s|\mathbf{x}),$$

where the term inside the integral is the probability that the sample path moves from $b(s|\mathbf{x})$ at time s to any level above $b(t|\mathbf{x})$ at time $t > s$. It is assumed that the underlying stochastic process has stationary independent increments.

In the example of Brownian motion, this equation takes the form

$$\Phi\left[-\frac{b(t|\mathbf{x})}{\sqrt{t}}\right] = \int_0^t \Phi\left[-\frac{b(t|\mathbf{x}) - b(s|\mathbf{x})}{\sqrt{t-s}}\right] dF(s|\mathbf{x}), \qquad [5.4]$$

(see Whitmore [WHI 86a]). Using the PH distribution function

$$F(s|\mathbf{x}) = 1 - \exp(-\exp(\boldsymbol{\beta}'\mathbf{x})H_0(s)),$$

gives an integral equation that can be solved numerically for $b(t|\mathbf{x})$ for specific values of x. A second example is given of the gamma process; this requires more extensive analysis.

5.3.3. *Estimation*

First, suppose that a PH model has been fitted to the data and it appears, on the basis of the available diagnostic tests, to be adequate. These regression results do not, however, provide any insight into the risk mechanism behind the pattern shown by the hazard function. Substituting the standard results from the PH output - the estimated regression coefficients $\hat{\beta}$ and the estimated cumulative baseline hazard $\hat{H}_0(t)$ - into the FHT model gives

$$S_{PH}(t|\mathbf{x}) = \exp\left[-\exp(\hat{\beta}'\mathbf{x})\hat{H}_0(t)\right],$$

and now this can be substituted for the true survival function for selected covariate vectors in order to estimate the corresponding family of process time functions or boundary functions.

Alternatively, estimation can be carried out without imposing the PH property. Simply take the usual Kaplan-Meier estimates $S_{KM}(t|\mathbf{x})$ of the survival function for selected $\mathbf{x}$, and substitute into the appropriate formulae, for example, into [5.3] (process time function) or [5.4] (boundary function) for a Brownian motion model. In the latter case, using the Kaplan-Meier estimator gives the following discrete approximation to the integral equation [5.4]:

$$\Phi\left[-\frac{b_j}{\sqrt{t_j}}\right] = \sum_{i=1}^{j} p_i \Phi\left[-\frac{b_j - b_i}{\sqrt{t_j - t_i}}\right],$$

where

$$\left(t_j, 1 - \sum_{i=1}^{j} p_i\right), \; j = 1, 2, \ldots,$$

is the Kaplan-Meier estimator at the ordered observed failure times $\{t_j; j = 1, \ldots, n\}$. The quantity Φ under the summation must be taken as 1/2 when $i = j$. Solving the above equation by numerical iteration for the b_j gives estimates of the boundary (see Whitmore [WHI 86a]). Extensions to continuous covariates are indicated by Lee and Whitmore, but not developed.

Estimation of a boundary function for the gamma process is stated to be straightforward in principle. It requires, however, the numerical evaluation of a double integral (Lee and Whitmore [LEE 10a]).

5.4. FHT and AFT models: theoretical connections

Following the analysis of the connections between FHT and PH models presented in the previous section, He and Whitmore [HE 13] developed the connections between FHT and AFT models. Again, the connection is made through the definition of a suitable running time.

One formulation of an AFT model is

$$Pr(T > t|A) = S(t|A) = S_0(t/A),$$

where A is the acceleration multiplier for an individual and $S_0(t)$ is a baseline survival function. In this expression, t is the calendar or clock time and $r = t/A$ is the transformed time. In the usual formulation of AFT models, A is a regression function of covariates $\mathbf{x}$, especially in the form $A = \exp(\boldsymbol{\beta}'\mathbf{x})$.

A slightly extended version of this is the more general statement of an AFT model as

$$Pr(T > t|\mathbf{x}) = S(t|\mathbf{x}) = S_0[r(t|\mathbf{x})],$$

where the non-decreasing function of calendar time $r(t|\mathbf{x})$ satisfies $r(0|\mathbf{x}) = 0, \forall \mathbf{x}$. Clearly, if S_0 is a first hitting time distribution in running time r for some baseline process and baseline boundary set, then the AFT model is also an FHT model. Although the AFT model is not an FHT model if $S_0(r)$ does not have this property, He and Whitmore [HE 13, p. 50] claim that their experience shows that "it is difficult to conceive of an AFT model that is scientifically meaningful that lies outside the [FHT] family". They go on to list a variety of FHT models that are also AFT models, starting with the Poisson process. The time until the first event in a Poisson process with baseline hazard rate λ_0 has the survival function

$$S_0(r) = \exp(-\lambda_0 r)$$

and with the simplest acceleration factor $A = \exp(\boldsymbol{\beta}'\mathbf{x})$, the survival function of the AFT model becomes

$$S(t|\mathbf{x}) = S_0(t/A) = \exp[-t\lambda_0 \exp(-\boldsymbol{\beta}'\mathbf{x})].$$

Similarly, if the time to the kth event is of interest, then $S_0(r)$ is given by an Erlang distribution and the AFT has a gamma error structure.

In the case of a Wiener process, $S_0(r)$ takes the inverse Gaussian form and with $r(t|\mathbf{x})$ as before, the AFT model has

$$S(t) = S_0[t\exp(-\boldsymbol{\beta}'\mathbf{x})].$$

Here, the running time function $r(t|\mathrm{x})$ characterizes the same AFT and FHT models. The FHT model would become broader than the AFT model if the boundary were made to depend on covariates.

Among the variants of the model presented by He and Whitmore is an exposure AFT model. This could be appropriate when the individual's exposure to harmful environmental factors varies between various periods. The varying exposure of railway workers to diesel exhaust fumes during different segments of their employment history is one example (see the papers by Lee *et al.* [LEE 04, LEE 09] in sections 6.2 and 6.3). Let

$$r(t|\mathbf{x}) = \sum_{j=1}^{J} \alpha_j(\mathbf{x}) t_j, \text{ where } t = \sum_{j=1}^{J} t_j,\ \alpha_j(\mathbf{x}) \geq 0, \text{ and } \alpha_1(\mathbf{x}) = 1,$$

where t_j is the duration of exposure to factor j or level j of one factor. The coefficients $\alpha_j(\mathbf{x})$ are the acceleration factors associated with exposure to each factor relative to a reference level. One special case of this model is

$$r(t|\mathbf{x}) = \begin{cases} t, & \text{if } t \leq c(\mathbf{x}) \\ c(\mathbf{x}) + [t - c(\mathbf{x})] \exp(\boldsymbol{\beta}'\mathbf{x}), & \text{if } t > c(\mathbf{x}) \end{cases}$$

where c is a change point, below which the factor has no effect but above which it causes acceleration.

6

Applications

6.1. Introduction

The number of published applications of first hitting time (FHT) regression models to the analysis of lifetime data is still limited. In fact, it is believed that this book contains all the applications of the model based on an underlying Wiener process that have appeared in the mainstream scientific literature up to the end of 2016.

As the model based on an underlying Wiener process has not yet entered into common use, routine applications can barely be found in the literature. Most papers that use this methodology are still partly fulfilling the role of attempting to raise the scientific community's awareness of the model, especially as an alternative to Cox's Proportional Hazards (PH) regression model. In addition, most present developments and extensions of the basic model. The papers on extensions that were presented in Chapter 4 emphasized methodology and used data chiefly for illustration, rather than carrying out in-depth analyses. In contrast, the papers that will be presented in this chapter place their emphasis on the substantive application, even if there are innovative methodological features.

The applications that have been presented in previous chapters will not be repeated here. Some of these were more in the nature of illustrations of the methodology or methodological developments rather than presentations of detailed analyses. For ease of reference, applications that were referred to earlier in the book are the following:

– *Kidney dialysis patients*: Lee and Whitmore (section 2.3; [LEE 10a]). A well-known set of data was used as an illustration of the FHT regression methodology in this article.

– *Exacerbations of chronic pulmonary obstructive disease*: Whitmore *et al.* (section 4.3; [WHI 12a]). The topic of chronic pulmonary obstructive disease appears again in section 6.7.

– *Mortality from malignant melanoma*: Pennell *et al.* (section 4.4; [PEN 10]) and Li and Lee (section 4.10; [LI 11]): the first paper introduced individual random effects into the model, fitted by Bayesian methods. This previously published set of data was used for illustration. The second was a methodological paper concerning the semi-parametric modeling of time-varying coefficients. The same data are also used for illustration of the *stthreg* package (section 3.3).

– *Repeated applications to drug treatment services by drug users*: Economou *et al.* (section 4.5; [ECO 15]). Different data from this database were used by Malefaki *et al.* as cited in the same section [MAL 15].

– *Analysis of low birth weights*: Whitmore and Su (section 4.7; [WHI 07]). This analysis is taken further in section 6.8.

– *Death or discharge of hospitalized elderly patients*: Horrocks and Thompson (section 4.6; [HOR 04]). The main purpose of this paper was to introduce a model with two or three absorbing barriers, in order to represent more than one possible outcome of the underlying process.

– *Sexually transmitted infections in adolescent women*: Yu *et al.* (section 4.8; [YU 09]). This paper introduced the semi-parametric modeling of covariate effects.

– *Treatments of soft tissue sarcoma*: Choi *et al.* (section 4.9; [CHO 14]). This paper also introduces semi-parametric modeling, in this case of the time scale. The presence of a cured fraction is an important feature of this model and the application.

– *Mortality after cardiac surgery*: Stogiannis *et al.* (section 5.2.1; [STO 11]). This case study compared results between FHT regression and Cox regression.

– *Lung cancer in a cohort of nurses*: Lee *et al.* (section 5.2.2; [LEE 10c]). This was another case study comparing FHT regression and Cox regression.

6.2. Lung cancer risk in railroad workers

The earliest detailed presentation of an application of the FHT regression model based on the Wiener process was published by Lee and her colleagues [LEE 04]. It uses the FHT model with an operational or process time scale, in order to assess mortality risk differentials of the working environments of 54,973 workers in the US railroad industry over periods of 10 to 20 years. The outcome variable is lung cancer mortality, and the focus of the investigation is on its relation to exposure to diesel exhaust fumes. Initially, a 20% sample had been used for model selection; this further analysis is based on the 80% confirmatory holdout sample.

In addition to providing a plausible representation of the mechanism of disease progression and death, the FHT model as used in this application presents the innovative feature of converting the real time scale to an operational time scale. Let the Wiener process $Y(r)$ represent health status and assume the operational time process $R(t)$. These two constituents, when combined, give the subordinated process $Y[R(t)]$ with parent process Y and directing process R.

Take the operational time process $R(t)$ to be

$$R(t) = \sum_{j=1}^{J} \int_0^t \alpha_j I_j(u) du = \sum_{j=1}^{J} \alpha_j A_j(t),$$

where $A_j(t)$ is the total duration of employment in job category $j = 1, \ldots, J$ up to time t. The positive coefficients $\alpha_j(t)$ – which are to be estimated – represent the relative exposures to diesel exhaust per unit of calendar time within each job category, setting $\alpha_J = 1$ for reference category J, which is taken to be retirement. The log likelihood is

$$\ell(\boldsymbol{\alpha}, \boldsymbol{\beta}, \boldsymbol{\gamma}) = \sum_{i=1}^{n} \{\delta_i \ln f(r_i|\mu_i, y_{0i}) + (1 - \delta_i) \ln S(r_i|\mu_i, y_{0i})\},$$

where $\boldsymbol{\beta}$ denotes the regression coefficients for μ, γ the regression coefficients for $\ln y_0$ and δ_i a censoring indicator. Inferences from the fitted model can be based on the usual asymptotic theory concerning maximum likelihood estimation. In addition – exploiting the very large sample size – the authors partitioned the full dataset into 25 subsets of equal size and refitted the model to each subset separately. For each parameter, the set of 25 independent estimates can be taken as the basis for inference based on the normal distribution by appeal to the central limit theorem.

A potential extension to the model, which is proposed but not developed in the paper, is to allow a random intercept in the regression functions for each subject. This would allow for further heterogeneity among subjects who have the same baseline characteristics and job exposures. The introduction of random intercepts into the standard model that uses only the calendar time scale [PEN 10] was described in section 4.4.

The same covariates – age (at the start of the study in 1959) and job category (in 1959; two dummy variables representing three categories) – were included in the linear predictors for both $\ln y_0$ and μ. Three coefficients α_j are to be estimated. P-values for parameter estimates (tested vs. zero) were generally smaller from the asymptotic analysis than from the complementary analysis based on 25 subsets, supporting the authors' caution in looking at the analysis based on the latter. Indeed, one coefficient seems to "lose significance" (at the conventional 5% level) in the subsets-based analysis.

Among the key results, it can be observed that the estimated value of μ is positive for all subjects, leading to the presence of long-term survivors (as regards death from lung cancer; other causes were not considered). All three estimates of α_j are < 1. Recall that the coefficient of the reference category, retirement, is set to 1. Thus, operational time progresses slower for someone in employment than someone who is retired. This could indicate a "healthy worker effect" - the phenomenon that good health is what allows a person to remain in employment and therefore creates the impression that employment, despite its stresses, is associated with better health. Concerning age (in 1959), its regression coefficient is significant ($P < 0.001$) for both $\ln y_0$ and μ. For the former, the coefficient is negative, implying that the initial health state declines with age, as seems logical. For the latter, however, the coefficient is positive, which corresponds to a stronger drift away from the boundary for older subjects. This apparent conflict is not discussed in the article, but the explanation for this finding could again lie in the healthy worker effect.

6.3. Lung cancer risk in railroad workers: a case-control study

The data in the analysis of lung cancer risk in railroad workers described above [LEE 04] were drawn from a retrospective cohort study. A further set of data on the same subjects were available from a case-control study and these data have also been re-analyzed by the same authors by application of FHT regression [LEE 09].

The data concern 1,256 workers who died of lung cancer in 1981–82, taken from the records held by the board responsible for managing the retirement system for railroad workers. For each of these cases, controls were selected at random from the same database, from among those whose birth date fell within 30 months of the case's birth and death date within 31 days (also there was mention of neither cancer on the death certificate, nor death from suicide, accident or unknown cause). In this way, a total of 2,385 controls were obtained; 90% of subjects had two controls, others had only one; 1,814 of the controls had died of circulatory disease and 577 of other causes.

The concept of the model is the same as has already been described for the 2004 analysis, with a disease-progression time scale related to the calendar time. However, it is necessary to take account of the case-control structure of the data in the likelihood. Because the underlying Wiener process is a time-reversible stochastic process, the (backward-looking) case-control study is equivalent to the (forward-looking) representation that would apply for a cohort study. Hence, the log likelihood can be written, similarly to the previous section, as

$$\ell(\boldsymbol{\alpha}, \boldsymbol{\beta}, \boldsymbol{\gamma}) = \sum_{i=1}^{n_1} \ln f\left(r_i|\mu_i, y_{0i}\right) + \sum_{i=n_1+1}^{n_1+n_2} \ln S\left(r_i|\mu_i, y_{0i}\right),$$

where there are n_1 cases and n_2 controls. No censoring indicator is required, because each of the cases (labeled $i = 1, \ldots, n_1$) contributes an observed survival time and each control (labeled from $n_1 + 1$ up to $n_1 + n_2$) contributes a survival time that is right censored for death from lung cancer (because death from another cause intervened).

Three analyses were conducted along these lines: (i) lung cancer deaths; (ii) cardiovascular deaths; (iii) other causes of death apart from lung cancer and cardiovascular disease. The covariates for $\ln y_0$ and μ were the following: smoking (ascertained from the next of kin – 17% missing data); engineer employment category (i.e. actual involvement in train operation – drivers, etc.); exposure to asbestos. All were significant for both parameters in each analysis except for the asbestos covariate in the analysis with other causes as cases. Concerning the estimates of the parameters of the underlying Wiener process, it can be noted that, in every analysis, the estimated coefficients of the same covariate always had opposite signs for $\ln y_0$ and μ. For example, for lung cancer deaths, engineers had a higher initial health status but a steeper decline. The authors discuss the implications of this feature in terms of the mean survival time $E(T) = y_0/|\mu|$, but do not provide an explanation of why it arose. One possible reason is that initial health status is higher because of the "healthy worker" effect – good health is demanded for this occupation – but there is higher exposure to risk, hence the faster drift.

The authors use modified Kaplan-Meier plots to aid their presentation of the results of the analyses. There are three (analyses) by eight (subgroups defined by combinations of engineer/smoking/asbestos) plots. Innovatively, the time scale is the operational time scale, not the calendar time as conventionally used (in other words, the Kaplan-Meier estimate is constructed non-parametrically as usual, except that "time" is that estimated from the model). For each outcome variable, there is a distinct clustering of the curves related to engineers, clearly separated from those for non-engineers (which show some degree of clustering with respect to smoking). This is not revealed at all in the conventional Kaplan-Meier plots against calendar time. Furthermore, the Kaplan-Meier survival curves for different subgroups cross over. This invalidates the use of proportional hazards modeling for these data.

The effects of employment are given by the coefficients α estimated for four categories relative to the reference category of retirement ($\alpha = 1$): years of life before taking up railroad employment; short exposure (number of years, up to 10); longer exposure (number of years beyond 10); years employed on railroad without exposure. We will take the analysis of lung cancer deaths as an illustration. For the first category – before entering railroad employment – the estimate of $\ln \alpha_1$ is -2.78791, so $\hat{\alpha}_1 = 0.062$; the process in this stage runs very slowly compared to retirement, at only 6% of the rate. This is expected because when people first enter this category they are young and have no exposure to risk. In the category of railroad employed but unexposed, $\hat{\alpha}_2 = 0.254$, again as expected. For exposure up to 10 years, the estimate $\hat{\alpha}_3 = 0.716$ relative to retirement continues to show a healthy worker effect (although this was not true for the other two analyses). Finally, for exposure beyond 10 years, $\hat{\alpha}_4 = 1.092$ (significantly greater than one) indicates the disappearance of the healthy worker effect, and the disease progresses at the same rate as in retirement or even faster.

6.4. Occupational exposure to asbestos

Like the study of lung cancer risk in railroad workers described in the preceding section, the study by Chambaz *et al.* [CHA 14] is a case-control study of lung cancer in relation to occupational exposure to a carcinogen. In this case, the carcinogen is asbestos. The study included 860 cases and 901 controls. The underlying model is the Wiener process, producing the inverse Gaussian distribution of lifetimes.

The effect of occupational exposure to asbestos is represented by accelerating the time scale of the process. Exposure is in 27 categories (plus one more for no exposure) obtained from the combination of three levels of probability of exposure, three levels of frequency and three levels of intensity. Although the details of the model are not the same as those of the model used by Lee *et al.* in the preceding section, the authors comment that their acceleration function is "very close to a linear combination of the times spent in each job category" [CHA 14, p.331] with coefficients that depend on the job categories, and therefore in this respect is "classical" – meaning, similar to the Lee and Whitmore formulation. However, the combination of measures of probability, frequency and intensity of exposure is original.

6.5. Return to work after limb injury

Hou and colleagues [HOU 08] carried out a conventional Cox PH regression in order to analyze the time until returning to work after injuries to the upper or lower limbs that required hospitalization, among people who were in full-time employment before their injury (which was not necessarily work related). These data, from Taiwan, included 154 cases, and predictors were measured at baseline. However, Kaplan-Meier curves in two important subgroups did not show the increasing separation that we expect to observe when the PH assumption holds.

This defect of the original analysis is recognized in a subsequent article by Hou and colleagues in which FHT regression based on an underlying Wiener process is used in place of Cox's model [HOU 16]. In the FHT framework, it is hypothesized that the patient's impairment can be quantified by a latent health index. This second analysis uses a much larger set of data than the first analysis, consisting of 1,124 cases, 75% of which returned to work within two years during the course of a prospective study with two-year follow-up. The nine predictors (all categorical), recorded at baseline, were entered into the model for both parameters of the Wiener process and backward elimination was used to remove predictors that were not significant at $P = 0.01$. This resulted in four predictors in common between the two parameters and one – labeled self-efficacy of return to work, that is, the patient's subjective probability of returning to work within one month – unique to the starting level parameter. From the signs of the coefficients in Table 2 of the article, holding a part time job was associated with a lower starting level (implying a tendency for a quicker return to work) but a greater positive value of the drift coefficient, hence a slower return to work. This discrepancy in signs did not arise for any other covariate.

The selection of variables in this analysis has some curious features. Although the underlying latent variable is described as measuring impairment, only two of the covariates appear to be directly relevant to the type or severity of the injury: location (upper limb or lower limb or other part of the body) and high (e.g. traffic accident) versus low energy injury. The latter was not significant. The other covariates – age, education, etc. – have at best an indirect relation to the degree of impairment, through their possible correlation with the severity of injury. A case could be made for including some of these covariates only in the regression for the drift parameter.

6.6. An FHT mixture model for a randomized clinical trial with switching

In the analysis of low birth weights in section 4.7, one of the models considered was a mixture of two inverse Gaussian distributions corresponding to different underlying processes. Lee *et al.* [LEE 08] also proposed using a mixture model to analyze data from a randomized clinical trial comparing a new therapy for multiple myeloma to an established one in 603 subjects. The design of the trial included switching patients from one drug to the other depending on his or her response. Because of this feature, a composite time scale is used, as we have seen in several other applications in which the process may proceed at different rates according to current conditions (see sections 4.9, 6.2, 6.3 and 6.4). Thus, if t_1 is the subject's calendar time on the primary therapy and t_2 the calendar time on the alternate therapy, the composite time is

$$r = \alpha t_1 + t_2,$$

where α is the rate of progression on the primary therapy relative to the alternate. Note that t_2 may be zero if in fact the subject remained on the primary therapy throughout the trial.

Simple Kaplan-Meier plots for one primary treatment against the other seem to point to the existence of an initial period of low risk which increases but then falls back again after roughly one year. This suggests that a mixture model may be required to represent the times of the event. The subsequent modeling continues to indicate the presence of a mixture after taking account of covariates. Thus, the analysis is based on the survival function

$$S(r) = pS_1(r) + (1-p)S_2(r)$$

in terms of composite time r. The covariates considered for inclusion are: the treatment to which the patient was randomized; whether disease progression was observed under primary therapy; the interaction between the first two covariates; whether the subject had received any treatments previously; the baseline level of beta-2 microglobulin; age at baseline.

Under this model, the sample log likelihood is

$$\ell(\boldsymbol{\theta}) = \sum_{i=1}^{n_1} \ln f(r_i|\boldsymbol{\theta}) + \sum_{i=n_1+1}^{n_1+n_2} \ln S(r_i|\boldsymbol{\theta}),$$

where n_1 participants labeled $i = 1, \ldots, n_1$ died and n_2 participants labeled $i = n_1 + 1, \ldots, n_1 + n_2$ survived to the end of the trial, thereby contributing a right-censored survival time. The parameters y_{0j} $(j = 1, 2)$ representing initial health status in the two components of the model are taken to depend on the three covariates that are known at baseline: age, beta-2 microglobulin level and previous treatment. The drift parameters μ_j $(j = 1, 2)$ are modeled as depending on the primary therapy, occurrence of disease progression and their interaction, because these are the variables that define the treatment received in the course of the trial. The mixing proportion p and the composite time weighting parameter α do not depend on covariates in the initial model.

The estimate of p obtained from maximization of the likelihood is 0.14 (95% confidence interval 0.09–0.21), which clearly supports the use of two components. It is noteworthy that the mean of μ_1 is positive under the new treatment, with or without disease progression. This indicates the presence of a cured fraction, because positive drift implies a non-zero probability of never reaching the boundary. The authors state that these, admittedly small, probabilities offer a slight hope in the context of a disease which hitherto had been seen as uniformly fatal.

The point estimate of the parameter α is 0.080. This value that is much less than 1 shows that latent health status declines slowly under the primary therapy until the switch to the alternate therapy. This result should be expected because the switch will not occur unless the disease has begun to accelerate substantially. Including the primary treatment in a regression function for $\ln \alpha$ led to an insignificant regression coefficient.

The above analysis is cited by Tamborrino *et al.* [TAM 15] who adopted a different approach to analyzing the impact of an intervention on the underlying Wiener process, namely modeling it as changing the drift parameter. Thus, their analysis revolves around an examination of pairs (S, R) where S is the time elapsed before intervention and R is the time from intervention to the event (crossing the threshold) or right censoring. The application is to data from a clinical trial in lung cancer patients, who were randomized to receive either standard or a test chemotherapy. S is the time from diagnosis to randomization and R is the survival time from then onwards. Note that truncation presents a difficulty that must be considered: subjects must survive long enough to reach intervention. However, this point had to be ignored in the application as the information was lacking. In a conventional analysis of the same data, Kalbfleisch and Prentice looked only at the time R, using S as one of the covariates [KAL 02].

One essential assumption of the analysis by Tamborrino *et al.* is that the timing of intervention is independent of the underlying process. The authors admit that this is a strong assumption which is probably not fulfilled in many cases, especially in medical contexts.

6.7. Recurrent exacerbations in COPD

A study carried out by Aaron and colleagues [AAR 10] investigates the timing of exacerbations of disease in patients suffering from chronic obstructive pulmonary disease (COPD), an incurable condition characterized by poor airflow leading to shortness of breath and other symptoms. It is said to affect almost 5% of the world's population. How can these recurrent events (exacerbations) be modeled?

The study tries out two approaches to the problem, using data from 449 patients who were taking part in a randomized clinical trial that included three treatment groups. One approach uses a model based on a Poisson process and the other an FHT model based on an underlying Wiener process. Interest in comparing these mechanisms arises because previous research indicates that exacerbations are most commonly caused by tracheo-bronchial infection, for which competing explanations of recurrence are either (1) persistence of an established infection that tends to resurface or (2) random acquisition of a new infection. These are represented by the Wiener and Poisson models, respectively (it is also possible that both processes are operating at the same time, or one in some patients and the other in others). See section 4.3 for another approach to the same data.

The underlying variable in the Wiener model is the patient's respiratory health. A stable interval inbetween exacerbations starts with an initial health level $y_0 > 0$ and ends when the health level falls to zero for the first time. The parameters of the Wiener process are regression functions of the patient's characteristics and environment, treatments and interventions received during the stable interval. Alternatively, in the Poisson model, exacerbations are regarded as a succession of events in a Poisson process, with a rate λ that could depend on the sequence number of exacerbation (it is a matter of observation that intervals tend to become shorter) or alternatively may be fixed for each patient but vary between patients. Two further variants of the models are considered. In the first, the unit of analysis is the stable interval (similar to the gaps between recurrent events in section 4.3 or to the way of writing the lifetime as a sequence of intervals for time-varying covariates in section 4.2). The predictors are the treatment group and the interval sequence number (a set of dummy variables).

Poisson model. In the analysis of stable intervals, it is necessary to include a probability of no exacerbation (some patients may not have one for a substantial number of years), so the survival function in the first stable interval is a modified exponential distribution:

$$P(S_{(1)} > s) = p + (1 - p)\exp(-\lambda s),$$

where $S_{(1)}$ is the length of this interval and p is the probability of never having an exacerbation. The regression models for $\ln(\lambda)$ and logit(p) are

$$\ln(\lambda) = \boldsymbol{\beta}'\mathbf{x}, \ \ \text{logit}(p) = \boldsymbol{\gamma}'\mathbf{x}\,.$$

The estimated probability of no exacerbation is $\hat{p} = 28\%$. The three treatment groups do not differ in a likelihood ratio test. The Poisson rate increases steadily with sequence number, in accordance with observation.

In the patient-level analysis, let the occurrence rate of exacerbations vary among patients according to a gamma distribution. The number of exacerbations during a particular patient's total observation time follows a Poisson distribution, but the overall distribution is negative binomial, with parameters depending on the length of follow-up and on covariates. Tests of goodness of fit were less satisfactory for this model than for the first variant, primarily because fewer patients than expected had large numbers of exacerbations.

Wiener model. The only notable difference from the Poisson model in the analysis of stable intervals arises from the need to include a delay t_1 in entry to the study (because the study protocol permitted randomization not less than 28 days after the last exacerbation), modeled by $\ln(t_1) = \boldsymbol{\zeta}'\mathbf{x}$ for the first interval. Thus, t_1 must be estimated in

$$P\left(S_{(1)} > s + t_1 | S_{(1)} > t_1\right),$$

and the regression models for the parameters of the Wiener process are

$$\mu = \boldsymbol{\beta}'\mathbf{x}, \ \ \ln y_0 = \boldsymbol{\gamma}'\mathbf{x}, \ \ \ln(t_1) = \boldsymbol{\zeta}'\mathbf{x}\,.$$

In the patient-level analysis, the parameters of the Wiener model are assumed to be the same for all patients. Despite this assumption of homogeneity of the population of patients, the fit of the model was adequate.

Comparing the Poisson and Wiener models in terms of the goodness of their fit to the data, the Wiener model was judged to be slightly preferable.

6.7.1. *COPD in lung cancer*

A further application of the Wiener process model in the context of the study of COPD – although with limited statistical interest – can be found in a recent analysis of survival following diagnosis of lung cancer, carried out by Ytterstad *et al.* [YTT 16]. In this study, the presence of COPD at baseline was included among the covariates. The primary analysis is a Cox proportional hazards regression. Modeling based on an underlying Wiener process (i.e. FHT regression) is performed only as a secondary analysis in order to "identify whether a covariate was of one of two types: 1) a measure of how far the cancer had advanced; or 2) a causal influence on the development of the disease" [YTT 16, p. 627]. The results of the FHT regression are not presented in detail; instead, the authors merely comment on which covariates were identified as being associated with survival in the FHT regression, in the course of their more extensive discussion of the results of what they call the "traditional" Cox regression analysis. This parallel use of two models that have different assumptions is unusual and rather hard to justify except for expository purposes as in section 5.2. No doubt the authors wished to use the Cox model for the ease of presenting its results in the familiar form of hazard ratios.

6.8. Normalcy and discrepancy indexes for birth weight and gestational age

The analysis of birth weights, which was instigated by Whitmore and Su [WHI 07] and is described in section 4.7, is taken further by Whitmore *et al.* [WHI 12b]. We recall that the earlier paper proposed that the birth outcome is determined by a stochastic process representing fetal development. Birth occurs when the process reaches a birthing boundary. Although fetal development is described by a two-dimensional process because both birth weight and gestational age are important variables, Whitmore and Su focused on birth weight in the original paper. The subsequent paper considers the joint outcome.

Let $U(r)$ denote fetal weight at fetal age r and $D(r)$ an unspecified measure of the physiological development of the fetus, with $U(0) = D(0) = 0$. Birth occurs when development $D(r)$ reaches the fixed boundary b for the first time, resulting in gestational age A at birth and birth weight $W = U(A)$. Assuming a Wiener process as usual, then the stopping time follows the inverse Gaussian distribution, using the parameterization

$$f(t|b,\mu) = \frac{b}{(2\pi t^3)^{1/2}} \exp\left[-\frac{(b-\mu t)^2}{2t}\right],$$

which is appropriate for a process with drift μ starting from zero and absorbed at boundary $b > 0$.

While Whitmore and Su – and also earlier analyses cited in that paper – suggested a two-component mixture of processes leading to "normal" and "abnormal" births, the present model includes three components in order to achieve adequate fit, with mixing proportions p_i, drift parameters μ_i and boundaries b_i, $i = 1, 2, 3$. Each birthing process falls into one of these three categories and is governed by the corresponding Wiener process. Interest focuses on two indexes describing the birth. These indexes are constructed from the actual birth weight W and a birth weight W_A imputed from the gestational age A under a power law relationship $W_A = \kappa A^\alpha$, where the estimates of the power law parameters κ and α are derived from a large sample of births in the USA.

The birth normalcy index (BNI) is the geometric mean $BNI = \sqrt{WW_A}$, which indicates whether birth weight and gestational age, viewed together, are abnormal.

The birth discrepancy index (BDI), indicating inconsistency between birth weight and gestational age, is defined as $BDI = \sqrt{W/W_A}$.

The overall pdf of BNI in the population of births is

$$f(BNI) = p_1 f_1(BNI) + p_2 f_2(BNI) + p_3 f_3(BNI),$$

where $f_i(BNI)$ denotes the pdf of BNI in birth category i. Empirically, a single distribution is found adequate to represent BDI and this is taken as a non-central t-distribution. For both indexes, the fitted distribution function is indistinguishable from the empirical distribution in the population.

The three components of the mixture are interpreted as follows. The largest category (with $\hat{p}_1 = 0.775$) represents healthy or "normal" births, with the highest mean value of the BNI. The smallest category, with $\hat{p}_3 = 0.027$, represents premature births. The boundary b_3 is much closer to zero than for the other categories. The remaining category lies in between the other two with respect to every parameter – drift, boundary and mixing proportion ($\hat{p}_2 = 0.198$). It has a substantial overlap with the healthy births and is labeled as an "at risk" category that includes moderately premature births, and also births at normal weight and gestational age that have other physiological problems. Much more extensive discussion than this brief summary can be found in the paper, which also introduces age, race and mode of delivery as covariates into the modeling of the three inverse Gaussian components [WHI 12b].

6.9. Hip fractures

The analyses of data on hip fractures described here refer to a prospective study of 9,704 women, 632 of whom suffered a hip fracture during the course of the study. An earlier analysis by Chapurlat *et al.* [CHA 03] examined the time from the initial fracture to a second one, among the subset of women who had at least one hip fracture (53 suffered a second hip fracture). Cox PH regression was applied. It is stated that the covariate "estrogen" failed a PH test, so a stratified analysis was conducted with this covariate as stratification factor (see section 1.15). Although this solves the problem of the violation of the PH assumption, it means that it is impossible to determine the relationship between estrogen use and the occurrence of a second hip fracture.

He *et al.* [HE 15] took the same dataset and analyzed the time to first fracture and separately the time from first to second fracture, using an FHT model based on an underlying Wiener process. The assumed underlying variable expresses skeletal strength. However, in addition, the model assumed that there is an independent stochastic process of random external shocks to the skeleton; for example, these might be caused by falls and other physical traumas. If the size of a shock experienced at time t exceeds skeletal strength at that time, then a fracture occurs. The authors name this the FHT degradation-shock model.

First, assume that skeletal strength decays from an initial level y_0, following an exponential trend at a constant rate $\lambda < 0$,

$$y(t) = y_0 \exp(\lambda t) .$$

(The authors comment that a development of the model would replace this deterministic decay by a random process.)

The random external shocks consist of a dense sequence of independent values from some distribution. The probability that all the shocks in a time interval of length δt are smaller than ν in size is $G(\nu)^{\delta t}$, where G is the distribution of maximum shock size per unit time. The person avoids hip fracture in that interval if the maximum shock is less than his or her current skeletal strength $y(t)$, from which it follows that the probability $S(t)$ of avoiding fracture throughout the interval $(0, t]$ is given by

$$\ln S(t) = \ln P(T > t) = \int_0^t \ln G\left[y(\tau)\right] d\tau \, . \qquad [6.1]$$

In order to proceed, suppose that $\ln G(\nu)$ has the reciprocal power form

$$\ln G(\nu) = -\left(\frac{\alpha}{\nu}\right)^{\beta}, \; \alpha > 0, \; \beta > 0,$$

which is the Fréchet extreme value distribution. Substituting this expression and the exponential distribution of $y(t)$ into [6.1] above gives

$$\ln S(t) = \ln P(T > t) = \frac{1}{\beta\lambda}\left(\frac{\alpha}{y_0}\right)^{\beta}\left[\exp(-\beta\lambda t) - 1\right] .$$

It can be seen from the form of this function that only two of the model's four parameters are estimable.

Special cases arise for $\lambda > 0$, giving the following non-zero probability of the individual never having a fracture

$$\ln P(T = \infty) = -\frac{1}{\beta\lambda}\left(\frac{\alpha}{y_0}\right)^{\beta}, \; \lambda > 0$$

and for $\lambda = 0$, giving the exponential survival function $S(t) = \exp(-\zeta t)$ where $\zeta = (\alpha/y_0)^{\beta}$.

The hazard function $h(t)$ for this degradation-shock model is

$$\ln h(t) = \beta\left[\ln(\alpha) - \ln(y_0) - \lambda t\right],$$

which is a decreasing function of time for positive λ, increasing for negative λ and constant for $\lambda = 0$.

In the further application of the model, baseline covariates are introduced by means of a logarithmic link function for y_0 and an identity link for λ. As noted above, the parameters α and β cannot be estimated separately and are set to 1 without loss of generality. Nineteen covariates are available. The covariates that are to be retained in the model are selected by backward elimination. In the analysis for the time to first fracture of the hip, eleven covariates remain in the model for initial skeletal strength y_0 and three for the degradation rate λ. The familiar problem of "opposite signs" arises for one covariate, estrogen, which has a positive coefficient for skeletal strength and negative for degradation rate λ. This implies that greater initial skeletal strength is associated with the current use of estrogen, but the subsequent decline is faster. Two explanations of this are offered. One is simply that the beneficial effect of estrogen declines with its continued use. The second is that some subjects who are on estrogen at baseline subsequently discontinue it because of the health risks posed by its long-term use, thereby raising their risks of hip fracture.

In the analysis for the time from first to second fracture, three covariates remain significant for initial skeletal strength and one (marginally) for degradation rate. The estimated values of λ are positive for all except 15 subjects; these are the ones with the highest values of the body mass index. This indicates that obese patients experience deterioration in skeletal strength at this stage. Body mass index was not significant for the first fracture. As noted above, $\lambda > 0$ gives a defective distribution of time to event, with a non-zero cured fraction or immunes. Fitting the model without covariates gives an estimate of the immune proportion of 91.4%; thus, few individuals go on to suffer a second fracture. This might be because of therapy or changed diet to improve bone strength, or behavioral changes to avoid risks.

Model checking includes the examination of martingale residuals $r_i = f_i - e_i$, where f_i is a dummy variable for fracture (1 = yes/0 = no) and e_i is "an unbiased estimate of f_i calculated from the model for the given covariates of patient i". These are plotted against each covariate or quantity of interest. The horizontal axis is the rank of the subject on this quantity (e.g. estimated initial skeletal strength in the example shown in the paper). On the vertical axis are plotted the cumulative counts of the actual fracture events f_i

and of the estimates e_i. This plot indicates both the calibration of the fitted model and its predictive power. Closeness of the two curves indicates good calibration. Specifically, the distance between them equals the cumulative sum of the martingale residuals. The closeness of curves to the top left-hand corner of the plot indicates strong predictive power.

6.10. Annual risk of death in cystic fibrosis

The background to cystic fibrosis (CF) is set out in Chin *et al.* [CHI 15]. It is the commonest genetic disease in Caucasian populations, recorded in 1/3,000 births. Although at one time, the life expectancy of people born with CF was only six months, modern medicine enables them to reach adulthood. However, the median age at death even in such an advanced country as Canada in 2013 was still only 35.1 years. Lung transplantation may be an option to extend the survival of the sickest CF patients. But even this drastic treatment is not a cure and it is associated with high risks of complications, perioperative morbidity and mortality; therefore, the selection of patients for transplantation is not a straightforward matter. Consequently, "an accurate understanding of a CF patient's prognosis and health trajectory is critical for appropriate patient selection and timing of transplantation" [CHI 15, p. 244].

The study by Aaron *et al.* [AAR 15] concentrates on estimating cystic fibrosis patients' risk of death in between scheduled annual visits to the clinic. Apart from anything else, the results of this analysis should be valuable for deciding on the timing of referral for lung transplant: patients with high risk of not surviving the year need to be referred now. The FHT model based on an underlying Wiener process is assumed, with the latent variable representing a CF health index. The model is similar to the one used for osteoporotic hip fractures described in the preceding section 6.9. Physical shocks in the model for hip fractures correspond to exacerbations of disease in the present application. Thus, the frequency of occurrence of exacerbations and their magnitude are governed by the Fréchet distribution. For the purpose of estimation, each patient's history is decomposed into a sequence of conditionally independent intervals between annual visits (i.e. conditionally independent, given baseline covariates for that interval). This assumes the Markovian property as discussed in section 4.2 [LEE 10b]. All the intervals, except possibly the last one, are right-censored observations of one year's duration. The estimates of the initial chronic health level y_0 and the exacerbation shape parameter β are based on 39,646 records from 3,794 patients.

Significant covariates (including interaction terms between the individual variables) were identified by backward elimination based on P-values. Eight covariates were retained in the model for y_0 and nine for β, but with only three in common. Two covariates - the presence of pseudomonas infection and being underweight (body mass index below 18.5) – have "opposite signs"; both have the expected negative coefficients for the starting level but unexpected positive coefficients for the drift parameter. This is discussed in some detail in terms of the hazard rate. For example, if two patients have identical covariates except that one is underweight but the other not, then the underweight patient has a higher mortality rate if the health level $\ln y_0$ is below 5.4. Thus, being underweight plays a role when the patient's health is already poor. A similar interpretation is given for the presence of pseudomonas [AAR 15, p. 1341].

Model checking in this analysis is carried out by constructing a plot for each covariate as described in the preceding section 6.9. The horizontal axis gives the record number of the case when ordered by the value of the covariate under consideration. Two curves are plotted: one is the cumulative count of deaths and the other is the cumulative sum of estimated death probabilities. If the two curves are close, this indicates good calibration of the regression model with respect to this covariate, and how close the curves are to the upper left-hand corner of the graph is indicative of the predictive value of the covariate. Another check is to compare actual and estimated deaths in quintiles of the estimated death probabilities (after some shrinkage of the probabilities).

Furthermore, the very large sample size permits the validation of the model on a 20% hold-out sample after fitting the model to an 80% training sample. In fact, this was carried out 100 times, partitioning the data randomly, and the results were averaged.

6.11. Disease resistance in cows

Saebø *et al.* [SAE 05] use a Wiener process to model the development of mastitis in cows, based on the records of 36,178 Norwegian cows. Three features regarding the structure of the model differentiate this work from other applications. First, a shift parameter ϕ is introduced so that it is $T - \phi$ that follows the inverse Gaussian distribution, and not simply a time T

measured from a known origin. This is because it is regarded as unlikely that the disease process actually starts at the chosen time origin (which is set at 31 days before calving). Second, it is believed that mastitis is a disease with many associated risk factors; therefore, a competing risks model should be adopted in order to represent the different causes. If these causes correspond to J independent latent processes, then the total hazard function at time t is the sum of the process-specific hazards that have been initiated by that time:

$$h(t|\boldsymbol{\theta}) = \sum_{j=1}^{J} h(t|\boldsymbol{\theta}_j) I(t > \phi_j),$$

where the parameter vector $\boldsymbol{\theta}_j$ includes, in addition to the regression coefficients for the parameters of the inverse Gaussian distribution in process j, the initiation time ϕ_j of that process. Third, random effects are included in the drift parameters μ_j. This is done because information is available on the 245 sires of these cows. Moreover, from Kaplan-Meier estimates produced separately for the daughters of each sire, it can be seen that there is substantial between-sire variability. In principle, random effects could be included in the initial levels as well.

The model for the drift parameter of the daughters of sire k in process j is

$$\mu_{jk} = \xi_j + s_{jk},$$

where ξ_j is a general mean and s_{jk} is the process-specific effect of sire k. It is assumed that $\mathbf{s}_j = (s_{j1}, \ldots, s_{j245})'$ is distributed as $N_{245}(\mathbf{0}, \sigma_j^2\mathbf{R})$, where the 245×245 known matrix $\mathbf{R}$ expresses the dependence structure of $\mathbf{s}_j$. It is constructed on the basis of the genetic relationships between individuals (information on ancestors of the sires was also available).

Combining these three features, the conditional likelihood function is

$$L(\mathbf{c}, \boldsymbol{\xi}, \boldsymbol{\phi}|\mathbf{t}, \mathbf{s}) = \prod_{i=1}^{n} \left[\sum_{j=1}^{J} \left\{ h\left(t_i|\xi_j, c_j, \phi_j, \mathbf{s}_{(i)j}\right) I\left(t_i > \phi_j\right) \right\}^{\delta_i} \right.$$
$$\left. \times \prod_{j=1}^{J} S\left(t_i|\xi_j, c_j, \phi_j, \mathbf{s}_{(i)j}\right)^{I(t_i > \phi_j)} \right],$$

where c is the starting level that we have denoted elsewhere as y_0.

A Bayesian approach using Markov chain Monte Carlo methods is adopted, with diffuse prior distributions for the parameter to be estimated. The number of latent processes was set at $J = 2$ based on physiological considerations, but alternatively the appropriate number could be decided upon by fitting the model for various values of J and choosing the best according to some criterion.

Because the time origin $t = 0$ is set arbitrarily to 31 days before calving, the estimated time shift parameter $\hat{\phi}_1 = 30.4$ for the first process indicates that it commences just the day before calving. On the other hand, $\hat{\phi}_2 = -16.4$, so the second process begins about 7 weeks earlier. The estimated values of the other parameters imply that both processes trend away from the boundary and consequently mastitis is not certain to occur. Combining the estimates of evading mastitis in both processes, the probability of not developing mastitis at all (the cured fraction) is

$$S(\infty|\boldsymbol{\theta}_1)S(\infty|\boldsymbol{\theta}_2) = \{1 - \exp(2\hat{c}_1\hat{\mu}_1)\}\{1 - \exp(2\hat{c}_2\hat{\mu}_2)\}$$

with a calculated value of 0.689 overall. For individual sires, the corresponding probability ranges from 0.552 to 0.769.

The authors' arguments in favor of the FHT approach include the suggestion that non-statisticians may well find the idea of a genetically dependent drift towards disease easier to understand than the genetically dependent hazard functions that a PH model would require.

6.12. Balka, Desmond and McNicholas: an application of their cure rate models

In section 2.4, it was mentioned that Balka *et al.* [BAL 09] introduced a variety of models for data including a cured fraction or long-term survivors based on the defective inverse Gaussian distribution. They illustrated some of these models on data obtained from Ichida *et al.*'s investigation of the effect of two disinfectants on the time to occurrence of *Staphylococcus aureus* infection in burn patients [ICH 93]. This serious infection – leading to increased length of stay in hospital and even causing death – is common but does not occur in all patients. The original analysis used Cox's PH regression model with time-varying covariates. It did not allow for a cured fraction. The re-analysis uses only the two baseline variables: type of disinfectant (two types; one dummy variable) and percentage of body surface burned (quantitative variable). Three related models based on an underlying Wiener process are considered. The first takes the inverse Gaussian distribution with drift parameter depending linearly on the two covariates. The second is a mixture model, composed of susceptibles to whom this inverse Gaussian distribution applies and an immune proportion. The probability of immunity (cure, long-term survival free of infection) of patient i is modeled by a logistic regression on the two covariates:

$$\ln\left(\frac{p_i}{1-p_i}\right) = \lambda_0 + \lambda_1 x_{1i} + \lambda_2 x_{2i}.$$

As we have commented earlier, this model partly loses the interpretation of the event as an outcome of the underlying process, because lack of susceptibility does not arise from the process. The third model is a two-barrier model, in which the upper absorbing barrier is the event of interest (infection) and the lower absorbing barrier is the acquisition of immunity (see also section 4.6).

The second and third models both have seven parameters, compared to the first model's four. All the three models have similar values of maximized log likelihood, therefore the AIC favors the simplest model, the first (Table 6.1). A log-normal mixture model was a little worse (AIC = 498.2) and a Weibull mixture worse still (AIC = 504.8). However, the estimated cure rate, also shown in Table 6.1, was quite different between the models despite the similarity of the fits.

		Estimated cure rate if 70% burned	
Model	AIC	Type 1	Type 2
Inverse Gaussian	490.85	0	0.40
IG cure rate	496.33	0.11	0.55
Two-barrier	493.69	0.18	0.54

Table 6.1. *Fit of different models considered by Balka* et al., *with estimated cure rates for two types of disinfectant (based on Tables 1 and 2 of [BAL 09])*

6.13. Progression of cervical dilation

The final illustrative application by Mulatya *et al.* refers to the modeling of a marker of progress towards the event of interest (see section 2.11), rather than the occurrence of the event itself, and thus is close in spirit to applications in the reliability literature where, for example, the increase in damage to a material might be measured rather than the actual time to failure through this damage. However, the authors of this biomedical example claim that it differs from reliability applications in that the intrinsic variability is much greater than is found in reliability data, and that multiple thresholds must be considered [MUL 16]. The context is the study of the progression of the dilation of the cervix in women in labor. The practical importance of the study is that if the dilation has failed to advance to a certain threshold in a given time, it may be appropriate to consider alternatives to a natural birth. The precise objective of the analysis is to fit an underlying Wiener process in order to estimate the distribution of transition times from one threshold to another, for example, from 3 cm dilation to 4 cm. This is accomplished by estimating, within a conditional linear FHT model, the probability

$$F_k\left[t|t_0, y_0, \boldsymbol{\theta}\right] = P\left\{\max_{u\in[t_0,t_0+t)} Y^*(u) \geq y_0 + k|Y^*(t_0) = y_0, \boldsymbol{\theta}(y_0)\right\}$$

that the process reaches $y_0 + k$ by time $t_0 + t$ given that it was at y_0 at time t_0. The underlying model is

$$Y_{ij} = Y_{i,j-1} + \beta_{Y_{i,j-1}}\left(t_{ij} - t_{i,j-1}\right) + \sigma B\left(t_{ij} - t_{i,j-1}\right),$$

where β is the drift, σ^2 is a variance parameter and the notation $B(t)$ denotes standard Brownian motion with $B(t) \sim N(0,t)$. Subscript i denotes the individual, j the observation on that individual and the parameter vector $\boldsymbol{\theta}$ is (β, σ). The model assumes linearity from one observation to the next, which is more plausible if the observations are close together.

An approach that has previously been applied to the same problem uses techniques for doubly interval-censored data required because the times at which lower and upper limits of the interval in question are reached are not known exactly but only known to lie in intervals. A criticism of this is that intermediate measurements are discarded. Mulatya *et al.* compare their longitudinal FHT regression model to the doubly interval-censored analysis on a sample of 2,000 births. Estimates of mean or median transition times were not greatly different between the two methods, but higher percentiles differed substantially. For example, the 95th percentile of time from 7 to 8 cm dilation was 5.45 hours (95% confidence interval 4.95–5.96) from the new model, but only 2.52 hours (1.76–3.27) from the old one. Such large differences could have a big impact on obstetrical practice.

Bibliography

[AAL 78] AALEN O.O., "Nonparametric inference for a family of counting processes", *Annals of Statistics*, vol. 6, no. 4, pp. 701–726, 1978.

[AAL 89] AALEN O.O., "A linear regression model for the analysis of life times", *Statistics in Medicine*, vol. 8, no. 8, pp. 907–925, 1989.

[AAL 94] AALEN O.O., "Effects of frailty in survival analysis", *Statistical Methods in Medical Research*, vol. 3, no. 3, pp. 227–243, 1994.

[AAL 01] AALEN O.O., GJESSING H.K., "Understanding the shape of the hazard rate: a process point of view", *Statistical Science*, vol. 16, no. 1, pp. 1–22, 2001.

[AAL 04] AALEN O.O., GJESSING H.K., "Survival models based on the Ornstein–Uhlenbeck process", *Lifetime Data Analysis*, vol. 10, no. 4, pp. 407–423, 2004.

[AAL 08] AALEN O.O., BORGAN Ø., GJESSING H.K., *Survival and Event History Analysis: A Process Point of View*, Springer, New York, 2008.

[AAR 10] AARON S.D., RAMSAY T., VANDEMHEEN K. *et al.*, "A threshold regression model for recurrent exacerbations in chronic obstructive pulmonary disease", *Journal of Clinical Epidemiology*, vol. 63, no. 12, pp. 1324–1331, 2010.

[AAR 15] AARON S.D., STEPHENSON A.L., CAMERON D.W. *et al.*, "A statistical model to predict one-year risk of death in patients with cystic fibrosis", *Journal of Clinical Epidemiology*, vol. 68, no. 11, pp. 1336–1345, 2015.

[ABD 75] ABDEL-HAMEED M., "A gamma wear process", *IEEE Transactions on Reliability*, vol. R-24, no. 2, pp. 152–153, 1975.

[AHE 16] A'HERN R.P., "Restricted mean survival time: an obligatory end point for time-to-event analysis in cancer trials?", *Journal of Clinical Oncology*, vol. 34, no. 28, pp. 3474–3476, 2016.

[ALT 95] ALTMAN D.G., DE STAVOLA B.L., LOVE S.B. *et al.*, "Review of survival analyses published in cancer journals", *British Journal of Cancer*, vol. 72, no. 2, pp. 511–518, 1995.

[ARG 09] ARGYROPOULOS C., CHANG C.-C.H., PLANTINGA L. *et al.*, "Considerations in the statistical analysis of hemodialysis patient survival", *Journal of the American Society of Nephrology*, vol. 20, no. 9, pp. 2034–2043, 2009.

[BAG 99] BAGDONAVIČIUS V.B., NIKULIN M.S., "Generalized proportional hazards model based on modified partial likelihood", *Lifetime Data Analysis*, vol. 5, no. 4, pp. 329–350, 1999.

[BAG 05] BAGDONAVIČIUS V.B., HAFDI M.A., HIMDI K.E. *et al.*, "Statistical analysis of the generalized linear proportional hazards model", *Journal of Mathematical Sciences*, vol. 127, no. 1, pp. 1673–1681, 2005.

[BAL 09] BALKA J., DESMOND A.F., MCNICHOLAS P.D., "Review and implementation of cure models based on first hitting times for Wiener processes", *Lifetime Data Analysis*, vol. 15, no. 2, pp. 147–176, 2009.

[BAL 11] BALKA J., DESMOND A.F., MCNICHOLAS P.D., "Bayesian and likelihood inference for cure rates based on defective inverse Gaussian regression models", *Journal of Applied Statistics*, vol. 38, no. 1, pp. 127–144, 2011.

[BAR 73] BARTHOLOMEW D.J., *Stochastic Models for Social Processes*, 2nd edition, John Wiley & Sons, Chichester, 1973.

[BAR 94] BARNETT V., LEWIS T., *Outliers in Statistical Data*, John Wiley & Sons, Chichester, 1994.

[BAR 99] BARR D.R., SHERRILL E.T., "Mean and variances of truncated normal distributions", *The American Statistician*, vol. 53, no. 4, pp. 357–361, 1999.

[BEN 86] BENDELL A., WALLEY M., WIGHTMAN D.W. *et al.*, "Proportional hazards modelling in reliability analysis – an application to brake discs on high speed trains", *Quality and Reliability Engineering International*, vol. 2, no. 1, pp. 45–52, 1986.

[BEN 91] BENDELL A., WIGHTMAN D.W., WALKER E.V., "Applying proportional hazards modelling in reliability", *Reliability Engineering and System Safety*, vol. 34, no. 1, pp. 35–53, 1991.

[BEN 83a] BENNETT S., "Analysis of survival data by the proportional odds model", *Statistics in Medicine*, vol. 2, no. 2, pp. 273–277, 1983.

[BEN 83b] BENNETT S., "Log-logistic regression models for survival data", *Applied Statistics*, vol. 32, no. 2, pp. 165–171, 1983.

[BHA 82] BHATTACHARYYA G.K., FRIES A., "Fatigue failure models – Birnbaum–Saunders vs. inverse Gaussian", *IEEE Transactions on Reliability*, vol. R-31, no. 5, pp. 439–440, 1982.

[BIR 69] BIRNBAUM Z.W., SAUNDERS S.C., "A new family of life distributions", *Journal of Applied Probability*, vol. 6, no. 2, pp. 319–327, 1969.

[BOA 49] BOAG J.W., "Maximum likelihood estimates of the proportion of patients cured by cancer therapy", *Journal of the Royal Statistical Society, Series B*, vol. 11, no. 1, pp. 15–53, 1949.

[BUR 16a] BURKE K., MACKENZIE G., "Multi-parameter regression survival modeling: an alternative to proportional hazards", *Biometrics*, 2016.

[BUR 16b] BURKE K., "Package mpr", available at: https://cran.r-project.org/web/packages/mpr/mpr.pdf, 2016.

[CAN 92] CANTOR A.B., SHUSTER J.J., "Parametric versus non-parametric methods for estimating cure rates based on censored survival data", *Statistics in Medicine*, vol. 11, no. 7, pp. 931–937, 1992.

[CAR 04] CARONI C., "Diagnostics for Cox's proportional hazards model", in NIKULIN M.S., BALAKRISHNAN N., MESBAH M. *et al.* (eds.), *Parametric and Semiparametric Models with Applications to Reliability, Survival and Quality of Life*, Birkhauser, Boston, 2004.

[CAR 10] CARONI C., "Testing for the Marshall–Olkin extended form of the Weibull distribution", *Statistical Papers*, vol. 51, no. 2, pp. 325–336, 2010.

[CAR 11a] CARONI C., "Cox regression for interval censored and current status data", *International Journal of Applied Mathematics and Statistics*, vol. 24, no. SI-11A, pp. 125–132, 2011.

[CAR 11b] CARONI C., "Graduation and attrition of engineering students in Greece", *European Journal of Engineering Education*, vol. 36, no. 1, pp. 63–74, 2011.

[CAR 14a] CARONI C., "Applicability of the Weibull distribution and its extensions in the biosciences", in OLIVEIRA T., BIEBLER K.-E., OLIVEIRA A. *et al.* (eds.), *Statistical and Biometrical Challenges, Theory and Applications*, Shaker Verlag, Aachen, 2014.

[CAR 14b] CARTER E.M., POTTS H.W.W., "Predicting length of stay from an electronic patient record system: a primary total knee replacement example", *BMC Medical Informatics and Decision Making*, vol. 14, no. 26, pp. 1–13, 2014.

[CHA 14] CHAMBAZ A., CHOUDAT D., HUBER C. *et al.*, "Analysis of the effect of occupational exposure to asbestos based on threshold regression modeling of case-control data", *Biostatistics*, vol. 15, no. 2, pp. 327–340, 2014.

[CHA 03] CHAPURLAT R.D., BAUER D.C., NEVITT M. *et al.*, "Incidence and risk factors for a second hip fracture in elderly women. The Study of Osteoporotic Fractures", *Osteoporosis International*, vol. 14, no. 2, pp. 130–136, 2003.

[CHE 00] CHEN Y.Q., WANG M.-C., "Analysis of accelerated hazards models", *Journal of the American Statistical Association*, vol. 95, no. 450, pp. 608–618, 2000.

[CHE 01a] CHEN Y.Q., "Accelerated hazards regression model and its adequacy for censored survival data", *Biometrics*, vol. 57, no. 3, pp. 853–860, 2001.

[CHE 01b] CHEN Y.Q., JEWELL N.P., "On a general class of semiparametric hazards regression models", *Biometrika*, vol. 88, no. 3, pp. 687–702, 2001.

[CHE 05] CHEN Y.Q., CHENG S., "Semiparametric regression analysis of mean residual life with censored survival data", *Biometrika*, vol. 92, no. 1, pp. 19–29, 2005.

[CHE 12] CHEN Y.Q., HU N., CHENG S.-C. *et al.*, "Estimating regression parameters in an extended proportional odds model", *Journal of the American Statistical Association*, vol. 107, no. 497, pp. 318–330, 2012.

[CHE 89] CHENG R.C.H., STEPHENS M.A., "A goodness-of-fit test using Moran's statistic with estimated parameters", *Biometrika*, vol. 76, no. 2, pp. 385–392, 1989.

[CHH 77] CHHIKARA R.S., FOLKS J.L., "The inverse Gaussian distribution as a lifetime model", *Technometrics*, vol. 19, no. 4, pp. 461–468, 1977.

[CHH 89] CHHIKARA R.S., FOLKS J.L., *The Inverse Gaussian Distribution: Theory, Methodology and Applications*, Marcel Dekker, New York, 1989.

[CHI 15] CHIN M., EARLAM K., AARON S.D., "Survival in cystic fibrosis: trends, clinical factors, and prediction models", *Pediatric Allergy, Immunology, and Pulmonology*, vol. 28, no. 4, pp. 244–249, 2015.

[CHO 14] CHOI S., HUANG X., CORMIER J.N. *et al.*, "A semiparametric inverse-Gaussian model and inference for survival data with a cured proportion", *The Canadian Journal of Statistics*, vol. 42, no. 4, pp. 635–649, 2014.

[CHU 91] CHUNG C.-F., SCHMIDT P., WITTE A.D., "Survival analysis: a survey", *Journal of Quantitative Criminology*, vol. 7, no. 1, pp. 59–98, 1991.

[CIA 85] CIAMPI A., ETEZADI-AMOLI J., "A general model for testing the proportional hazards and the accelerated failure time hypotheses in the analysis of censored survival data with covariates", *Communications in Statistics – Theory and Methods*, vol. 14, no. 3, pp. 651–667, 1985.

[COL 14] COLLETT D., *Modelling Survival Data in Medical Research*, 3rd edition, Chapman and Hall, Boca Raton, 2014.

[COM 13] COMMENGES D., HEJBLUM B.P., "Evidence synthesis through a degradation model applied to myocardial infarction", *Lifetime Data Analysis*, vol. 19, no. 1, pp. 1–18, 2013.

[COO 82] COOK R.D., WEISBERG S., *Residuals and Influence in Regression*, Chapman and Hall, New York, 1982.

[COX 65] COX D.R., MILLER H.D., *The Theory of Stochastic Processes*, Methuen, London, 1965.

[COX 68] COX D.R., SNELL E.J., "A general definition of residuals", *Journal of the Royal Statistical Society, Series B*, vol. 30, no. 2, pp. 248–275, 1968.

[COX 72] COX D.R., "Regression models and life-tables", *Journal of the Royal Statistical Society, Series B*, vol. 34, no. 2, pp. 187–220, 1972.

[COX 74] COX D.R., HINKLEY D.V., *Theoretical Statistics*, Chapman and Hall, London, 1974.

[COX 75] COX D.R., "Partial likelihood", *Biometrika*, vol. 62, no. 2, pp. 269–276, 1975.

[COX 83] COX D.R., "A remark on censoring and surrogate response variables", *Journal of the Royal Statistical Society, Series B*, vol. 45, no. 3, pp. 391–393, 1983.

[COX 90] COX D.R., "Role of models in statistical analysis", *Statistical Science*, vol. 5, no. 2, pp. 169–174, 1990.

[COX 99] COX D.R., "Some remarks on failure-times, surrogate markers, degradation, wear, and the quality of life", *Lifetime Data Analysis*, vol. 5, no. 4, pp. 307–314, 1999.

[CRO 01] CROWDER M.J., *Classical Competing Risks*, Chapman and Hall, Boca Raton, 2001.

[DAL 85] DALE C.J., "Application of the proportional hazards model in the reliability field", *Reliability Engineering*, vol. 10, no. 1, pp. 1–14, 1985.

[DAU 03] DAUXOIS J.-Y., KIRMANI S.N.U.A., "Testing the proportional odds model under random censoring", *Biometrika*, vol. 90, no. 4, pp. 913–922, 2003.

[DAV 80] DAVIS A.S., "Use of the likelihood ratio test on the inverse Gaussian distribution", *The American Statistician*, vol. 34, no. 2, pp. 108–110, 1980.

[DAV 03] DAVISON A.C., *Statistical Models*, Cambridge University Press, Cambridge, 2003.

[DEM 77] DEMPSTER A.P., LAIRD N.M., RUBIN D.B., "Maximum likelihood from incomplete data via the EM algorithm", *Journal of the Royal Statistical Society, Series B*, vol. 39, no. 1, pp. 1–38, 1977.

[DEN 91] DENNIS B., MUNHOLLAND P.L., SCOTT J.M., "Estimation of growth and extinction parameters for endangered species", *Ecological Monographs*, vol. 61, no. 2, pp. 115–143, 1991.

[DER 05] DEROSE JR. J.J., TOUMPOULIS I.K., BALARAM S.K. *et al.*, "Preoperative prediction of long-term survival after coronary artery bypass grafting in patients with low left ventricular ejection fraction", *The Journal of Thoracic and Cardiovascular Surgery*, vol. 129, no. 2, pp. 314–321, 2005.

[DES 85] DESMOND A.F., "Stochastic models of failure in random environments", *The Canadian Journal of Statistics*, vol. 13, no. 2, pp. 171–183, 1985.

[DES 86] DESMOND A.F., "On the relationship between two fatigue-life models", *IEEE Transactions on Reliability*, vol. R-35, no. 2, pp. 167–169, 1986.

[DES 93] DESMOND A.F., CHAPMAN G.R., "Modelling task completion data with inverse Gaussian mixtures", *Applied Statistics*, vol. 42, no. 4, pp. 603–613, 1993.

[DIA 92] DIAMOND I.D., MCDONALD J.W., "Analysis of current-status data", in TRUSSELL J., HANKINSON R., TILTON J. (eds.), *Demographic Applications of Event History Analysis*, Clarendon Press, Oxford, 1992.

[DUC 00] DUCHESNE T., LAWLESS J., "Alternative time scales and failure time models", *Lifetime Data Analysis*, vol. 6, no. 2, pp. 157–179, 2000.

[EAT 77] EATON W.W., WHITMORE G.A., "Length of stay as a stochastic process: a general approach and application to hospitalization for schizophrenia", *Journal of Mathematical Sociology*, vol. 5, no. 2, pp. 273–292, 1977.

[EBE 01] EBERLY L.E., GRAMBSCH P., CONNETT J.E., "Comment on paper by Aalen and Gjessing", *Statistical Science*, vol. 16, no. 1, pp. 16–19, 2001.

[ECO 07] ECONOMOU P., CARONI C., "Parametric proportional odds frailty models", *Communications in Statistics-Simulation and Computation*, vol. 36, no. 6, pp. 1295–1307, 2007.

[ECO 15] ECONOMOU P., MALEFAKI S., CARONI C., "Bayesian threshold regression model with random effects for recurrent events", *Methodology and Computing in Applied Probability*, vol. 17, no. 4, pp. 871–898, 2015.

[EFR 77] EFRON B., "The efficiency of Cox's likelihood function for censored data", *Journal of the American Statistical Association*, vol. 72, no. 359, pp. 557–565, 1977.

[ELA 80] ELANDT-JOHNSON R.C., "Some prior and posterior distributions in survival analysis: a critical insight on relationships derived from cross-sectional data", *Journal of the Royal Statistical Society, Series B*, vol. 42, no. 1, pp. 96–106, 1980.

[ELS 90] ELSAYED E.A., CHAN C.K., "Estimation of thin-oxide reliability using proportional hazards models", *IEEE Transactions on Reliability*, vol. 39, no. 3, pp. 329–335, 1990.

[ERI 15] ERICH R., PENNELL M.L., "Ornstein–Uhlenbeck threshold regression for time-to-event data with and without a cure fraction", *Lifetime Data Analysis*, vol. 21, no. 1, pp. 1–19, 2015.

[ESC 92] ESCOBAR L.A., MEEKER W.Q., "Assessing influence in regression analysis with censored data", *Biometrics*, vol. 48, no. 2, pp. 507–528, 1992.

[ETE 87] ETEZADI-AMOLI J., CIAMPI A., "Extended hazard regression for censored survival data with covariates: a spline approximation for the baseline hazard function", *Biometrics*, vol. 43, no. 1, pp. 181–192, 1987.

[FAD 99] FADDY M.J., MCCLEAN S.I., "Analysing data on lengths of stay of hospital patients using phase-type distributions", *Applied Stochastic Models in Business and Industry*, vol. 15, no. 4, pp. 311–317, 1999.

[FAR 79] FAREWELL V.T., COX D.R., "A note on multiple time scales in life testing", *Applied Statistics*, vol. 28, no. 1, pp. 73–75, 1979.

[FAR 82] FAREWELL V.T., "The use of mixture models for the analysis of survival data with long-term survivors", *Biometrics*, vol. 38, no. 4, pp. 1041–1046, 1982.

[FAR 86] FAREWELL V.T., "Mixture models in survival analysis: are they worth the risk?", *The Canadian Journal of Statistics*, vol. 14, no. 3, pp. 257–262, 1986.

[FEL 86] FELLER W., *An Introduction to Probability Theory and Its Applications*, vol. 1, John Wiley & Sons, New York, 1986.

[FIS 99] FISHER L.D., LIN D.Y., "Time-dependent covariates in the Cox proportional-hazards regression model", *Annual Review of Public Health*, vol. 20, pp. 145–157, 1999.

[FOL 78] FOLKS J.L., CHHIKARA R.S., "The inverse Gaussian distribution and its statistical application – a review", *Journal of the Royal Statistical Society, Series B*, vol. 40, no. 3, pp. 263–289, 1978.

[FOR 95] FORD I., NORRIE J., AHMADI S., "Model inconsistency, illustrated by the Cox proportional hazards model", *Statistics in Medicine*, vol. 14, no. 8, pp. 735–746, 1995.

[FRE 08] FREEDMAN D.A., "Survival analysis: a primer", *The American Statistician*, vol. 62, no. 2, pp. 110–119, 2008.

[FRI 83] FRIES A., BHATTACHARYYA G.K., "Analysis of two-factor experiments under an inverse Gaussian model", *Journal of the American Statistical Association*, vol. 78, no. 384, pp. 820–826, 1983.

[GAI 84] GAIL M.H., WIEAND S., PIANTADOSI S., "Biased estimates of treatment effect in randomized experiments with nonlinear regressions and omitted covariates", *Biometrika*, vol. 71, no. 3, pp. 431–444, 1984.

[GER 01] GERDS T.A., SCHUMACHER M., "On functional misspecification of covariates in the Cox regression model", *Biometrika*, vol. 88, no. 2, pp. 572–580, 2001.

[GHI 05] GHITANY M.E., AL-HUSSAINI E.K., AL-JARALLAH R.A., "Marshall–Olkin extended Weibull distribution and its application to censored data", *Journal of Applied Statistics*, vol. 32, no. 10, pp. 1025–1034, 2005.

[GIE 98] GIESER P.W., CHANG M.N., RAO P.V. *et al.*, "Modelling cure rates using the Gompertz model with covariate information", *Statistics in Medicine*, vol. 17, no. 8, pp. 831–839, 1998.

[GJE 15] GJESSING H.K., "Package 'invGauss'", available at: https://cran.r-project.org/web/packages/invGauss/invGauss.pdf, 2015.

[GOE 17] GOEMAN J., MEIJER R., CHATUVEDI N. *et al.*, "Package 'penalized'", available at: https://cran.r-project.org/web/packages/penalized/penalized.pdf, 2017.

[GOR 10] GORJIAN N., MA L., MITTINTY M. *et al.*, "A review on degradation models in reliability analysis", in KIRITSIS D., EMMANOUILIDIS C., KORONIOS A. *et al.* (eds.), *Engineering Asset Lifecycle Management*, Springer-Verlag, London, 2010.

[GRA 99] GRAF E., SCHMOOR C., SAUERBREI W. *et al.*, "Assessment and comparison of prognostic classification schemes for survival data", *Statistics in Medicine*, vol. 18, nos. 17–18, pp. 2529–2545, 1999.

[GUP 07] GUPTA R.C., GUPTA R.D., "Proportional reversed hazard rate model and its applications", *Journal of Statistical Planning and Inference*, vol. 137, no. 11, pp. 3525–3536, 2007.

[HAR 07] HARDIN J.W., HILBE J.M., *Generalized Linear Models and Extensions*, 2nd edition, Stata Press, College Station, 2007.

[HAS 03] HASHEMI R., JACQMIN-GADDA H., COMMENGES D., "A latent process model for joint modeling of events and marker", *Lifetime Data Analysis*, vol. 9, no. 4, pp. 331–343, 2003.

[HE 13] HE X., WHITMORE G.A., "Connecting threshold regression and accelerated failure time models", in LEE M.-L.T., GAIL M., PFEIFFER R. *et al.* (eds.), *Risk Assessment and Evaluation of Predictions*, Springer, New York, 2013.

[HE 15] HE X., WHITMORE G.A., LOO G.Y. *et al.*, "A model for time to fracture with a shock stream superimposed on progressive degradation: the Study of Osteoporotic Fractures", *Statistics in Medicine*, vol. 34, no. 4, pp. 652–663, 2015.

[HOR 04] HORROCKS J., THOMPSON M.E., "Modeling event times with multiple outcomes using the Wiener process with drift", *Lifetime Data Analysis*, vol. 10, no. 1, pp. 29–49, 2004.

[HOU 92] HOUGAARD P., HARVALD B., HOLM N.V., "Measuring the similarities between the lifetimes of adult Danish twins born between 1881–1930", *Journal of the American Statistical Association*, vol. 87, no. 417, pp. 17–24, 1992.

[HOU 94] HOUGAARD P., MYGLEGAARD P., BORCH-JOHNSEN K., "Heterogeneity models of disease susceptibility, with application to diabetic nephropathy", *Biometrics*, vol. 50, no. 4, pp. 1178–1188, 1994.

[HOU 95] HOUGAARD P., "Frailty models for survival data", *Lifetime Data Analysis*, vol. 1, no. 3, pp. 255–273, 1995.

[HOU 99] HOUGAARD P., "Fundamentals of survival data", *Biometrics*, vol. 55, no. 1, pp. 13–22, 1999.

[HOU 08] HOU W.-H., TSAUO J.-Y., LIN C.-H. *et al.*, "Worker's compensation and return-to-work following orthopaedic injury to extremities", *Journal of Rehabilitation Medicine*, vol. 40, no. 6, pp. 440–445, 2008.

[HOU 16] HOU W.-H., CHUANG H.-Y., LEE M.-L.T., "A threshold regression model to predict return-to-work after traumatic limb injury", *Injury*, vol. 47, no. 2, pp. 483–489, 2016.

[HUT 02] HUTTON J.L., MONAGHAN P.F., "Choice of parametric accelerated life and proportional hazards models for survival data: asymptotic results", *Lifetime Data Analysis*, vol. 8, no. 4, pp. 375–393, 2002.

[HUZ 99] HUZURBAZAR A.V., "Flowgraph models for generalized phase type distributions having non-exponential waiting times", *Scandinavian Journal of Statistics*, vol. 26, no. 1, pp. 145–157, 1999.

[ICH 93] ICHIDA J.M., WASSELL J.T., KELLER M.D. *et al.*, "Evaluation of protocol change in burn-care management using the Cox proportional hazards model with time-dependent covariates", *Statistics in Medicine*, vol. 12, nos. 3–4, pp. 301–310, 1993.

[JEW 96] JEWELL N.P., KALBFLEISCH J.D., "Marker processes in survival analysis", *Lifetime Data Analysis*, vol. 2, no. 1, pp. 15–29, 1996.

[JEW 03] JEWELL N.P., VAN DER LAAN M.J., "Current status data: review, recent developments and open problems", *Handbook of Statistics*, vol. 23, pp. 625–642, 2003.

[KAL 89] KALBFLEISCH J.D., LAWLESS J.F., "Inference based on retrospective ascertainment: an analysis of the data on transfusion-related AIDS", *Journal of the American Statistical Association*, vol. 84, no. 406, pp. 360–372, 1989.

[KAL 02] KALBFLEISCH J.D., PRENTICE R.L., *The Statistical Analysis of Failure Time Data*, 2nd edition, John Wiley & Sons, Hoboken, 2002.

[KAY 02] KAY R., KINNERSLEY N., "On the use of the accelerated failure time model as an alternative to the proportional hazards model in the treatment of time to event data: a case study in influenza", *Drug Information Journal*, vol. 36, no. 3, pp. 571–579, 2002.

[KEI 97] KEIDING N., ANDERSEN P.K., KLEIN J.P., "The role of frailty models and accelerated failure time models in describing heterogeneity due to omitted covariates", *Statistics in Medicine*, vol. 16, no. 2, pp. 215–224, 1997.

[KIE 88] KIEFER N.M., "Economic duration data and hazard functions", *Journal of Economic Literature*, vol. 26, no. 2, pp. 646–679, 1988.

[KLE 03] KLEIN J.P., MOESCHBERGER M.L., *Survival Analysis: Techniques for Censored and Truncated Data*, 2nd edition, Springer-Verlag, New York, 2003.

[KOR 93] KORDONSKY K.B., GERTSBAKH I.B., "Choice of the best time scale for system reliability analysis", *European Journal of Operational Research*, vol. 65, no. 2, pp. 235–246, 1993.

[KOR 97] KORDONSKY K.B., GERTSBAKH I.B., "Multiple time scales and the lifetime coefficient of variation: engineering applications", *Lifetime Data Analysis*, vol. 2, no. 2, pp. 139–156, 1997.

[KRI 02] KRIVTSOV V.V., TANANKO D.E., DAVIS T.P., "Regression approach to tire reliability analysis", *Reliability Engineering and System Safety*, vol. 78, no. 3, pp. 267–273, 2002.

[KUM 94] KUMAR D., KLEFSJÖ B., "Proportional hazards model: a review", *Reliability Engineering and System Safety*, vol. 44, no. 2, pp. 177–188, 1994.

[KWO 03] KWONG G.P.S., HUTTON J.L., "Choice of parametric models in survival analysis: applications to monotherapy for epilepsy and cerebral palsy", *Applied Statistics*, vol. 52, no. 2, pp. 153–168, 2003.

[LAN 72] LANCASTER T., "A stochastic model for the duration of a strike", *Journal of the Royal Statistical Society, Series A*, vol. 135, no. 2, pp. 257–271, 1972.

[LAW 86] LAWLESS J.F., "A note on lifetime regression models", *Biometrika*, vol. 73, no. 2, pp. 509–512, 1986.

[LAW 03] LAWLESS J.F., *Statistical Models and Methods for Lifetime Data*, 2nd edition, John Wiley & Sons, Hoboken, 2003.

[LAW 04] LAWLESS J.F., CROWDER M., "Covariates and random effects in a gamma process model with application to degradation and failure", *Lifetime Data Analysis*, vol. 10, no. 3, pp. 213–227, 2004.

[LEE 00] LEE M.-L.T., DEGRUTTOLA V., SCHOENFELD D., "A model for markers and latent health status", *Journal of the Royal Statistical Society, Series B*, vol. 62, no. 4, pp. 747–762, 2000.

[LEE 02a] LEE M.-L.T., WHITMORE G.A., "Assumptions of a latent survival model", in HUBER-CAROL C., BALAKRISHNAN N., NIKULIN M.S. *et al.* (eds.), *Goodness-of-Fit Tests and Model Validity*, Birkhauser, Boston, 2002.

[LEE 02b] LEE M.-L.T., WHITMORE G.A., "A model relating quality of life to latent health status and survival", in MESBAH M., COLE B.C., LEE M.-L.T. (eds.), *Statistical Methods for Quality of Life Studies*, Springer, Boston, 2002.

[LEE 04] LEE M.-L.T., WHITMORE G.A., LADEN F. *et al.*, "Assessing lung cancer risk in railroad workers using a first hitting time regression model", *Environmetrics*, vol. 15, no. 5, pp. 501–512, 2004.

[LEE 06] LEE M.-L.T., WHITMORE G.A., "Threshold regression for survival analysis: modeling event times by a stochastic process reaching a boundary", *Statistical Science*, vol. 21, no. 4, pp. 501–513, 2006.

[LEE 08] LEE M.-L.T., CHANG M., WHITMORE G.A., "A threshold regression mixture model for assessing treatment efficacy in a multiple myeloma clinical trial", *Journal of Biopharmaceutical Statistics*, vol. 18, no. 6, pp. 1136–1149, 2008.

[LEE 09] LEE M.-L.T., WHITMORE G.A., LADEN F. *et al.*, "A case-control study relating railroad worker mortality to diesel exhaust exposure using a threshold regression model", *Journal of Statistical Planning and Inference*, vol. 139, no. 5, pp. 1633–1642, 2009.

[LEE 10a] LEE M.-L.T., WHITMORE G.A., "Proportional hazards and threshold regression: their theoretical and practical connections", *Lifetime Data Analysis*, vol. 16, no. 2, pp. 196–214, 2010.

[LEE 10b] LEE M.-L.T., WHITMORE G.A., ROSNER B., "Threshold regression for survival data with time-varying covariates", *Statistics in Medicine*, vol. 29, nos. 7–8, pp. 896–905, 2010.

[LEE 10c] LEE M.-L.T., WHITMORE G.A., ROSNER B., "Benefits of threshold regression: a case-study comparison with Cox proportional hazards regression", in RYKOV V.V., BALAKRISHNAN N., NIKULIN M.S. (eds.), *Mathematical and Statistical Models and Methods in Reliability: Applications to Medicine, Finance, and Quality Control*, Birkhauser, Boston, 2010.

[LI 11] LI J., LEE M.-L.T., "Analysis of failure time using threshold regression with semi-parametric varying coefficients", *Statistica Neerlandica*, vol. 65, no. 2, pp. 164–182, 2011.

[LIN 94] LIN D.Y., YING Z., "Semiparametric analysis of the additive risk model", *Biometrika*, vol. 81, no. 1, pp. 61–71, 1994.

[LIN 08] LINDQVIST B.H., SKOGSRUD G., "Modeling of dependent competing risks by first passage times of Wiener processes", *IIE Transactions*, vol. 41, no. 1, pp. 72–80, 2008.

[LON 89] LONGINI JR. I.M., CLARK W.S., BYERS R.H. *et al.*, "Statistical analysis of the stages of HIV infection using a Markov model", *Statistics in Medicine*, vol. 8, no. 7, pp. 831–843, 1989.

[LU 93] LU C.J., MEEKER W.Q., "Using degradation measures to estimate a time-to-failure distribution", *Technometrics*, vol. 35, no. 2, pp. 161–174, 1993.

[LU 96] LU C.J., MEEKER W.Q., ESCOBAR L.A., "A comparison of degradation and failure-time analysis methods for estimating a time-to-failure distribution", *Statistica Sinica*, vol. 6, no. 3, pp. 531–546, 1996.

[MAD 13] MADEIRA S., INFANTE P., DIDELET F., "Use of survival models in a refinery", *Revstat – Statistical Journal*, vol. 11, no. 1, pp. 45–65, 2013.

[MAG 94] MAGULURI G., ZHANG C.-H., "Estimation in the mean residual life regression model", *Journal of the Royal Statistical Society, Series B*, vol. 56, no. 3, pp. 477–489, 1994.

[MAL 96] MALLER R.A., ZHOU X., *Survival Analysis with Long-term Survivors*, John Wiley & Sons, Chichester, 1996.

[MAL 15] MALEFAKI S., ECONOMOU P., CARONI C., "Modelling times between events with a cured fraction using a first hitting time regression model with individual random effects", in KITSOS C.P., OLIVEIRA T.A., RIGAS A. *et al.* (eds.), *Theory and Practice of Risk Assessment*, Springer, New York, 2015.

[MAR 97] MARSHALL A.W., OLKIN I., "A new method for adding a parameter to a family of distributions with application to the exponential and Weibull families", *Biometrika*, vol. 84, no. 3, pp. 641–652, 1997.

[MAR 07] MARSHALL A.W., OLKIN I., *Life Distributions: Structure of Nonparametric, Semiparametric, and Parametric Families*, Springer, New York, 2007.

[MAR 13] MARSHALL A.H., ZENGA M., GIORDANO S., "Modelling students' length of stay at university using Coxian phase-type distributions", *International Journal of Statistics and Probability*, vol. 2, no. 1, pp. 73–89, 2013.

[MCC 80] MCCOOL J.I., "Confidence limits for Weibull regression with censored data", *IEEE Transactions on Reliability*, vol. R-29, no. 2, pp. 145–150, 1980.

[MCC 89] MCCULLAGH P., NELDER J.A., *Generalized Linear Models*, 2nd edition, Chapman and Hall, London, 1989.

[MCD 96] MCDONALD D.B., FITZPATRICK J.W., WOOLFENDEN G.E., "Actuarial senescence and demographic heterogeneity in the Florida Scrub Jay", *Ecology*, vol. 77, no. 8, pp. 2373–2381, 1996.

[MEE 98a] MEEKER W.Q., ESCOBAR L.A., *Statistical Methods for Reliability Data*, John Wiley & Sons, New York, 1998.

[MEE 98b] MEEKER W.Q., ESCOBAR L.A., LU C.J., "Accelerated degradation tests: modeling and analysis", *Technometrics*, vol. 40, no. 2, pp. 98–99, 1998.

[MUD 96] MUDHOLKAR G.S., HUTSON A.D., "The exponentiated Weibull family: some properties and a flood data application", *Communications in Statistics – Theory and Methods*, vol. 25, no. 12, pp. 3059–3083, 1996.

[MUL 16] MULATYA C.M., MCLAIN A.C., CAI B. *et al.*, "Estimating time to event characteristics via longitudinal threshold regression models – an application to cervical dilation progression", *Statistics in Medicine*, vol. 35, no. 24, pp. 4368–4379, 2016.

[NAK 75] NAKAGAWA T., OSAKI S., "The discrete Weibull distribution", *IEEE Transactions on Reliability*, vol. R-24, no. 5, pp. 300–301, 1975.

[NAR 03] NARDI A., SCHEMPER M., "Comparing Cox and parametric models in clinical studies", *Statistics in Medicine*, vol. 22, no. 23, pp. 3597–3610, 2003.

[NEL 72] NELSON W., "Graphical analysis of accelerated life test data with the inverse power law model", *IEEE Transactions on Reliability*, vol. R-21, no. 1, pp. 2–11, 1972.

[NEL 90] NELSON W., *Accelerated Testing: Statistical Models, Test Plans, and Data Analysis*, John Wiley & Sons, New York, 1990.

[NIK 10] NIKULIN M.S., LIMNIOS N., BALAKRISHNAN N. *et al.*, *Advances in Degradation Modelling: Applications in Reliability, Survival Analysis, and Finance*, Birkhauser, Boston, 2010.

[OAK 77] OAKES D., "The asymptotic information in censored survival data", *Biometrika*, vol. 64, no. 3, pp. 441–448, 1977.

[OAK 90] OAKES D., DASU T., "A note on residual life", *Biometrika*, vol. 77, no. 2, pp. 409–410, 1990.

[OAK 95] OAKES D., "Multiple time scales in survival analysis", *Lifetime Data Analysis*, vol. 1, no. 1, pp. 7–18, 1995.

[OAK 03] OAKES D., DASU T., "Inference for the proportional mean residual life model", *Institute of Mathematical Statistics Lecture Notes – Monograph Series*, vol. 43, pp. 105–116, 2003.

[OAK 13] OAKES D., "An introduction to survival models: in honor of Ross Prentice", *Lifetime Data Analysis*, vol. 19, no. 4, pp. 442–462, 2013.

[OLK 16] OLKIN I., "Life distributions: a brief discussion", *Communications in Statistics – Simulation and Computation*, vol. 45, no. 5, pp. 1489–1498, 2016.

[PAD 04] PADGETT W.J., TOMLINSON M.A., "Inference from accelerated degradation and failure data based on Gaussian process models", *Lifetime Data Analysis*, vol. 10, no. 2, pp. 191–206, 2004.

[PAR 05] PARK C., PADGETT W.J., "Accelerated degradation models for failure based on geometric Brownian motion and gamma processes", *Lifetime Data Analysis*, vol. 11, no. 4, pp. 511–527, 2005.

[PAT 06] PATEL K., KAY R., ROWELL L., "Comparing proportional hazards and accelerated failure time models: an application in influenza", *Pharmaceutical Statistics*, vol. 5, no. 3, pp. 213–224, 2006.

[PEN 01] PENG Y., DEAR K.B.G., CARRIERE K.C., "Testing for the presence of cured patients: a simulation study", *Statistics in Medicine*, vol. 20, no. 12, pp. 1783–1796, 2001.

[PEN 07] PENG L., HUANG Y., "Survival analysis with temporal covariate effects", *Biometrika*, vol. 94, no. 3, pp. 719–733, 2007.

[PEN 10] PENNELL M.L., WHITMORE G.A., LEE M.-L.T., "Bayesian random-effects threshold regression with application to survival data with nonproportional hazards", *Biostatistics*, vol. 11, no. 1, pp. 111–126, 2010.

[PET 99] PETTIT L.I., YOUNG K.D.S., "Bayesian analysis for inverse Gaussian lifetime data with measures of degradation", *Journal of Statistical Computation and Simulation*, vol. 63, no. 3, pp. 217–234, 1999.

[PIJ 91] PIJNENBURG M., "Additive hazards models in repairable systems reliability", *Reliability Engineering and System Safety*, vol. 31, no. 3, pp. 369–390, 1991.

[PRE 89] PRENTICE R.L., "Surrogate endpoints in clinical trials: definition and operational criteria", *Statistics in Medicine*, vol. 8, no. 4, pp. 431–440, 1989.

[REI 94] REID N., "A conversation with Sir David Cox", *Statistical Science*, vol. 9, no. 3, pp. 439–455, 1994.

[RIC 88] RICCIARDI L.M., SATO S., "First-passage-time density and moments of the Ornstein–Uhlenbeck process", *Journal of Applied Probability*, vol. 25, no. 1, pp. 43–57, 1988.

[RIE 91] RIECK J.R., NEDELMAN J.R., "A log-linear model for the Birnbaum–Saunders distribution", *Technometrics*, vol. 33, no. 1, pp. 51–60, 1991.

[RIG 05] RIGBY R.A., STASINOPOULOS D.M., "Generalized additive models for location, scale and shape", *Applied Statistics*, vol. 54, no. 3, pp. 507–554, 2005.

[SAE 05] SAEBØ S., ALMØY T., AASTVEIT A.H., "Disease resistance modelled as first-passage times of genetically dependent stochastic processes", *Applied Statistics*, vol. 54, no. 1, pp. 273–285, 2005.

[SAN 08] SANKARAN P.G., JAYAKUMAR K., "On proportional odds models", *Statistical Papers*, vol. 49, no. 4, pp. 779–789, 2008.

[SCH 79] SCHMEE J., HAHN G.J., "A simple method for regression analysis with censored data", *Technometrics*, vol. 21, no. 4, pp. 417–432, 1979.

[SCH 97] SCHMOOR C., SCHUMACHER M., "Effects of covariate omission and categorization when analysing randomized trials with the Cox model", *Statistics in Medicine*, vol. 16, no. 3, pp. 225–237, 1997.

[SCH 87] SCHUMACHER M., OLSCHEWSKI M., SCHMOOR C., "The impact of heterogeneity on the comparison of survival times", *Statistics in Medicine*, vol. 6, no. 7, pp. 773–784, 1987.

[SES 97] SESHADRI V., *The Inverse Gaussian Distribution*, Springer-Verlag, New York, 1997.

[SHY 99] SHYUR H.-J., ELSAYED E.A., LUXHOJ J.T., "A general hazard regression model for accelerated life testing", *Annals of Operations Research*, vol. 91, no. 0, pp. 263–280, 1999.

[SI 11] SI X.-S., WANG W., HU C.-H. *et al.*, "Remaining useful life estimation – A review on the statistical data driven approaches", *European Journal of Operational Research*, vol. 213, no. 1, pp. 1–14, 2011.

[SIN 95] SINGPURWALLA N.D., "Survival in dynamic environments", *Statistical Science*, vol. 10, no. 1, pp. 86–103, 1995.

[SLU 14] SLUD E.V., SUNTORNCHOST J., "Parametric survival densities from phase-type models", *Lifetime Data Analysis*, vol. 20, no. 3, pp. 459–480, 2014.

[SOL 84] SOLOMON P.J., "Effect of misspecification of regression models in the analysis of survival data", *Biometrika*, vol. 71, no. 2, pp. 291–298, 1984.

[STA 07] STASINOPOULOS D.M., RIGBY R.A., "Generalized additive models for location, scale and shape (GAMLSS) in R", *Journal of Statistical Software*, vol. 23, no. 7, pp. 1–46, 2007.

[STA 16] STASINOPOULOS D.M., RIGBY R.A., "Package 'gamlss.cens'", available at: https://cran.r-project.org/web/packages/gamlss.cens/gamlss.cens.pdf, 2016.

[STO 11] STOGIANNIS D., CARONI C., ANAGNOSTOPOULOS C.E. *et al.*, "Comparing first hitting time and proportional hazards regression models", *Journal of Applied Statistics*, vol. 38, no. 7, pp. 1483–1492, 2011.

[STO 12] STOGIANNIS D., CARONI C., "Tests for outliers in the inverse Gaussian distribution, with application to first hitting time models", *Journal of Statistical Computation and Simulation*, vol. 82, no. 1, pp. 73–80, 2012.

[STO 13] STOGIANNIS D., CARONI C., "Issues in fitting inverse Gaussian first hitting time regression models for lifetime data", *Communications in Statistics – Simulation and Computation*, vol. 42, no. 9, pp. 1948–1960, 2013.

[STO 16] STOGIANNIS D., CARONI C., Variable selection in inverse Gaussian first hitting time regression models, Unpublished Technical Report, National Technical University of Athens, 2016.

[STR 86] STRUTHERS C.A., KALBFLEISCH J.D., "Misspecified proportional hazard models", *Biometrika*, vol. 73, no. 2, pp. 363–369, 1986.

[SUN 06] SUN J., *The Statistical Analysis of Interval-censored Failure Time Data*, Springer, New York, 2006.

[SUN 15] SUN J., FENG Y., ZHAO H., "Simple estimation procedures for regression analysis of interval-censored failure time data under the proportional hazards model", *Lifetime Data Analysis*, vol. 21, no. 1, pp. 138–155, 2015.

[TAM 15] TAMBORRINO T., DITLEVSEN S., LANSKY P., "Parameter inference from hitting times for perturbed Brownian motion", *Lifetime Data Analysis*, vol. 21, no. 3, pp. 331–352, 2015.

[THE 00] THERNEAU T.M., GRAMBSCH P.M., *Modeling Survival Data: Extending the Cox Model*, Springer, New York, 2000.

[TIB 96] TIBSHIRANI R., "Regression shrinkage and selection via the lasso", *Journal of the Royal Statistical Society, Series B*, vol. 58, no. 1, pp. 267–288, 1996.

[TIB 97] TIBSHIRANI R., "The lasso method for variable selection in the Cox model", *Statistics in Medicine*, vol. 16, no. 4, pp. 385–395, 1997.

[TSO 03] TSODIKOV A.D., IBRAHIM J.G., YAKOVLEV A.Y., "Estimating cure rates from survival data: an alternative to two-component mixture models", *Journal of the American Statistical Association*, vol. 98, no. 464, pp. 1063–1078, 2003.

[TWE 45] TWEEDIE M.C.K., "Inverse statistical covariates", *Nature*, vol. 155, pp. 453–453, 1945.

[TWE 57] TWEEDIE M.C.K., "Statistical properties of inverse Gaussian distributions. I.", *Annals of Mathematical Statistics*, vol. 28, no. 2, pp. 362–377, 1957.

[UPA 96] UPADHYAY S.K., AGRAWAL R., SMITH A.F.M., "Bayesian analysis of inverse Gaussian non-linear regression by simulation", *Sankhyā: The Indian Journal of Statistics, Series B*, vol. 58, no. 3, pp. 363–378, 1996.

[VAN 09] VAN NOORTWIJK J.M., "A survey of the application of gamma processes in maintenance", *Reliability Engineering and System Safety*, vol. 94, no. 1, pp. 2–21, 2009.

[VAU 79] VAUPEL J.W., MANTON K.G., STALLARD E., "The impact of heterogeneity in individual frailty on the dynamics of mortality", *Demography*, vol. 16, no. 3, pp. 439–454, 1979.

[VAU 85] VAUPEL J.W., YASHIN A.I., "Heterogeneity's uses: some surprising effects of selection on population dynamics", *The American Statistician*, vol. 39, no. 3, pp. 176–185, 1985.

[WAN 07] WANG H., LENG C., "Unified LASSO estimation by least squares approximation", *Journal of the American Statistical Association*, vol. 102, no. 479, pp. 1039–1048, 2007.

[WAN 10] WANG X., XU D., "An inverse Gaussian process model for degradation data", *Technometrics*, vol. 52, no. 2, pp. 188–197, 2010.

[WAN 16] WANG L., MCMAHAN C.S., HUDGENS M.G. *et al.*, "A flexible, computationally efficient method for fitting the proportional hazards model to interval-censored data", *Biometrics*, vol. 72, no. 1, pp. 222–231, 2016.

[WHI 75] WHITMORE G.A., "The inverse Gaussian distribution as a model of hospital stay", *Health Services Research*, vol. 10, pp. 297–302, 1975.

[WHI 78] WHITMORE G.A., "Discussion of the paper by Professor Folks and Dr. Chhikara", *Journal of the Royal Statistical Society, Series B*, vol. 40, no. 3, pp. 285–286, 1978.

[WHI 79] WHITMORE G.A., "An inverse Gaussian model for labour turnover", *Journal of the Royal Statistical Society, Series A*, vol. 142, no. 4, pp. 468–478, 1979.

[WHI 83] WHITMORE G.A., "A regression method for censored inverse-Gaussian data", *The Canadian Journal of Statistics*, vol. 11, no. 4, pp. 305–315, 1983.

[WHI 86a] WHITMORE G.A., "First-passage-time models for duration data: regression structures and competing risks", *The Statistician*, vol. 35, no. 2, pp. 207–219, 1986.

[WHI 86b] WHITMORE G.A., "Normal-gamma mixtures of inverse Gaussian distributions", *Scandinavian Journal of Statistics*, vol. 13, no. 3, pp. 211–220, 1986.

[WHI 95] WHITMORE G.A., "Estimating degradation by a Wiener diffusion process subject to measurement error", *Lifetime Data Analysis*, vol. 1, no. 3, pp. 307–319, 1995.

[WHI 97] WHITMORE G.A., SCHENKELBERG F., "Modelling accelerated degradation data using Wiener diffusion with a time scale transformation", *Lifetime Data Analysis*, vol. 3, no. 1, pp. 27–45, 1997.

[WHI 98] WHITMORE G.A., CROWDER M.J., LAWLESS J.F., "Failure inference from a marker process based on a bivariate Wiener model", *Lifetime Data Analysis*, vol. 4, no. 3, pp. 229–251, 1998.

[WHI 07] WHITMORE G.A., SU Y., "Modeling low birth weights using threshold regression: results for U.S. birth data", *Lifetime Data Analysis*, vol. 13, no. 2, pp. 161–190, 2007.

[WHI 12a] WHITMORE G.A., RAMSAY T., AARON S.D., "Recurrent first hitting times in Wiener diffusion under several observation schemes", *Lifetime Data Analysis*, vol. 18, no. 2, pp. 157–176, 2012.

[WHI 12b] WHITMORE G.A., ZHANG G., LEE M.-L.T., "Constructing normalcy and discrepancy indexes for birth weight and gestational age using a threshold regression mixture model", *Biometrics*, vol. 68, no. 1, pp. 297–306, 2012.

[WU 12] WU L., LI H., "Variable selection for joint mean and dispersion models of the inverse Gaussian distribution", *Metrika*, vol. 75, no. 6, pp. 795–808, 2012.

[XIA 12] XIAO T., WHITMORE G.A., HE X. *et al.*, "Threshold regression for time-to-event analysis: the stthreg package", *The Stata Journal*, vol. 12, no. 2, pp. 257–283, 2012.

[XIA 15] XIAO T., WHITMORE G.A., HE X. *et al.*, "The R package threg to implement threshold regression models", *Journal of Statistical Software*, vol. 66, no. 8, pp. 1–16, 2015.

[XIE 13] XIE X., STRICKLER H.D., XUE X., "Additive hazard regression models: an application to the natural history of human papillomavirus", *Computational and Mathematical Models in Medicine*, vol. 2013, pp. 1–7, 2013.

[XU 11] XU R., MCNICHOLAS P.D., DESMOND A.F. *et al.*, "A first passage time model for long-term survivors with competing risks", *The International Journal of Biostatistics*, vol. 7, no. 1, pp. 1–15, 2011.

[YE 14] YE Z.-S., CHEN N., "The inverse Gaussian process as a degradation model", *Technometrics*, vol. 56, no. 3, pp. 302–311, 2014.

[YTT 16] YTTERSTAD E., MOE P.C., HJALMARSEN A., "COPD in primary lung cancer patients: prevalence and mortality", *International Journal of COPD*, vol. 11, pp. 625–636, 2016.

[YU 09] YU Z., TU W., LEE M.-L.T., "A semi-parametric threshold regression analysis of sexually transmitted infections in adolescent women", *Statistics in Medicine*, vol. 28, no. 24, pp. 3029–3042, 2009.

[ZEL 59] ZELEN M., "Factorial experiments in life testing", *Technometrics*, vol. 1, no. 3, pp. 269–288, 1959.

[ZHA 09] ZHANG J., PENG Y., "Crossing hazard functions in common survival models", *Statistics and Probability Letters*, vol. 79, no. 20, pp. 2124–2130, 2009.

[ZUC 06] ZUCKER D.M., YANG S., "Inference for a family of survival models encompassing the proportional hazards and proportional odds models", *Statistics in Medicine*, vol. 25, no. 6, pp. 995–1014, 2006.

Index

A, B, C

D, E, F

G, H, I

K, L, M

N, O, P, R

S, T, V, W

Other titles from

iSTE

in

Mathematics and Statistics

2017

CELANT Giorgio, BRONIATOWSKI Michel
Interpolation and Extrapolation Optimal Designs 2: Finite Dimensional General Models

D'AMICO Guglielmo, DI BIASE Giuseppe, JANSSEN Jacques, MANCA Raimondo
Semi-Markov Migration Models for Credit Risk (Stochastic Models for Insurance Set – Volume 1)

HARLAMOV Boris
Stochastic Analysis of Risk and Management (Stochastic Models in Survival Analysis and Reliability Set – Volume 2)

NIKULIN Mikhail, CHIMITOVA Ekaterina
Chi-squared Goodness-of-fit Tests for Censored Data (Stochastic Models in Survival Analysis and Reliability Set – Volume 3)

SIMON Jacques
Banach, Fréchet, Hilbert and Neumann Spaces (Analysis for PDEs Set – Volume 1)

2016

CELANT Giorgio, BRONIATOWSKI Michel
Interpolation and Extrapolation Optimal Designs 1: Polynomial Regression and Approximation Theory

CHIASSERINI Carla Fabiana, GRIBAUDO Marco, MANINI Daniele
Analytical Modeling of Wireless Communication Systems (Stochastic Models in Computer Science and Telecommunication Networks Set – Volume 1)

GOUDON Thierry
Mathematics for Modeling and Scientific Computing

KAHLE Waltraud, MERCIER Sophie, PAROISSIN Christian
Degradation Processes in Reliability
(Mathematial Models and Methods in Reliability Set – Volume 3)

KERN Michel
Numerical Methods for Inverse Problems

RYKOV Vladimir
Reliability of Engineering Systems and Technological Risks
(Stochastic Models in Survival Analysis and Reliability Set – Volume 1)

2015

DE SAPORTA Benoîte, DUFOUR François, ZHANG Huilong
Numerical Methods for Simulation and Optimization of Piecewise Deterministic Markov Processes

DEVOLDER Pierre, JANSSEN Jacques, MANCA Raimondo
Basic Stochastic Processes

LE GAT Yves
Recurrent Event Modeling Based on the Yule Process
(Mathematical Models and Methods in Reliability Set – Volume 2)

2014

COOKE Roger M., NIEBOER Daan, MISIEWICZ Jolanta
Fat-tailed Distributions: Data, Diagnostics and Dependence (Mathematical Models and Methods in Reliability Set – Volume 1)

MACKEVIČIUS Vigirdas
Integral and Measure: From Rather Simple to Rather Complex

PASCHOS Vangelis Th
Combinatorial Optimization – 3-volume series – 2nd edition
Concepts of Combinatorial Optimization / Concepts and Fundamentals – volume 1
Paradigms of Combinatorial Optimization – volume 2
Applications of Combinatorial Optimization – volume 3

2013

COUALLIER Vincent, GERVILLE-RÉACHE Léo, HUBER Catherine, LIMNIOS Nikolaos, MESBAH Mounir
Statistical Models and Methods for Reliability and Survival Analysis

JANSSEN Jacques, MANCA Oronzio, MANCA Raimondo
Applied Diffusion Processes from Engineering to Finance

SERICOLA Bruno
Markov Chains: Theory, Algorithms and Applications

2012

BOSQ Denis
Mathematical Statistics and Stochastic Processes

CHRISTENSEN Karl Bang, KREINER Svend, MESBAH Mounir
Rasch Models in Health

DEVOLDER Pierre, JANSSEN Jacques, MANCA Raimondo
Stochastic Methods for Pension Funds

2011

MACKEVIČIUS Vigirdas
Introduction to Stochastic Analysis: Integrals and Differential Equations

MAHJOUB Ridha
Recent Progress in Combinatorial Optimization – ISCO2010

RAYNAUD Hervé, ARROW Kenneth
Managerial Logic

2010

BAGDONAVIČIUS Vilijandas, KRUOPIS Julius, NIKULIN Mikhail
Nonparametric Tests for Censored Data

BAGDONAVIČIUS Vilijandas, KRUOPIS Julius, NIKULIN Mikhail
Nonparametric Tests for Complete Data

IOSIFESCU Marius *et al.*
Introduction to Stochastic Models

VASSILIOU PCG
Discrete-time Asset Pricing Models in Applied Stochastic Finance

2008

ANISIMOV Vladimir
Switching Processes in Queuing Models

FICHE Georges, HÉBUTERNE Gérard
Mathematics for Engineers

HUBER Catherine, LIMNIOS Nikolaos *et al.*
Mathematical Methods in Survival Analysis, Reliability and Quality of Life

JANSSEN Jacques, MANCA Raimondo, VOLPE Ernesto
Mathematical Finance

2007

HARLAMOV Boris
Continuous Semi-Markov Processes

2006

CLERC Maurice
Particle Swarm Optimization

Printed and bound by CPI Group (UK) Ltd, Croydon, CR0 4YY

13/09/2026

14961024-0002